A Handbook of Tropical Soil Biology

A Handbook of Tropical Soil Biology

Sampling and Characterization of Below-ground Biodiversity

Edited by Fatima M. S. Moreira, E. Jeroen Huising and David E. Bignell

earthscan
publishing for a sustainable future

London • Sterling, VA

First published by Earthscan in the UK and USA in 2008

Hardback ISBN: 978-1-84407-621-5
Paperback ISBN: 978-1-84407-593-5

Typeset by MapSet Ltd, Gateshead, UK
Printed and bound in the UK by Cromwell Press, Trowbridge, Wiltshire
Cover design by Edward Bignell and Susanne Harris

For a full list of publications please contact:

Earthscan
Dunstan House
14a St Cross Street
London, EC1N 8XA, UK
Tel: +44 (0)20 7841 1930
Fax: +44 (0)20 7242 1474
Email: earthinfo@earthscan.co.uk
Web: **www.earthscan.co.uk**

22883 Quicksilver Drive, Sterling, VA 20166-2012, USA

Earthscan publishes in association with the International Institute for Environment and Development

A catalogue record for this book is available from the British Library

Library of Congress Cataloging-in-Publication Data has been applied for

The paper used for this book is FSC-certified. FSC (the Forest Stewardship Council) is an international network to promote responsible management of the world's forests.

Mixed Sources
Product group from well-managed forests and other controlled sources
www.fsc.org Cert no. TT-COC-2082
© 1996 Forest Stewardship Council

This publication presents part of the findings of the international project 'Conservation and Sustainable Management of Below-ground Biodiversity' implemented in seven tropical countries – Brazil, Côte d'Ivoire, India, Indonesia, Kenya, Mexico and Uganda. This project is coordinated by the Tropical Soil Biology and Fertility Institute of CIAT (TSBF-CIAT) with co-financing from the Global Environment Facility (GEF), and implementation support from the United Nations Environment Programme (UNEP).

TSBF-CIAT
c/o ICRAF, United Nations Avenue, Gigiri
PO Box 30677-00100, Nairobi, Kenya
Web: www.ciat.cgiar.org/tsbf; www.bgbd.net

Contents

List of Plates, Figures and Tables

Plates

three layers (L1, L2 and L3), h) spore of *Paraglomus brasilianum* showing the spore wall structure formed by three layers (L1, L2 and L3) (notice that L2 is ornamented with minute ridges). Pictures of *Paraglomus occultum* and *P. brasilianum* are from http://invam.caf.wvu.edu.

Plate 6 Isolation of zoosporic fungi from environmental samples (Peronosporomycetes, Cytridiomycetes): a) soil sample with baits; b) water sample with baits; c) pure culture on bait; d) sporangia of *Phytophthora*; e) oogonium and antheridum of *Pythium*; f) *Pythium* liberating zoospores.

Plate 7 Soil washing technique for isolation of soil microfungi: a) pre-washing; b) sieves with different mesh; c) pre-washed soil; d) continued washing procedure; e) washed soil particles; f) plating on culture medium; g) plate with soil particles for incubation; h) fungal colonies on culture medium with growth retardants.

Plate 8 a) Larvae and pupae of fruit fly in the soil; b) MacPhail trap containing a food bait; c) common wing patterns with stripes; d) extrovertion of the aculeus; e) adult of fruit fly and detail of apical aculeus.

FIGURES

Tables

Contributors

Abreu, L. M., Instituto de Química, Universidade Federal de Minas Gerais, Belo Horizonte, MG, Brazil. lmabreu@gmail.com

Bagyaraj, J. D., Department of Agricultural Microbiology, University of Agricultural Sciences, GKVK Campus, Bangalore, India. dbagyaraj@vsnl.com

Bignell, D. E., Institute for Tropical Biology and Conservation, Universiti Malaysia Sabah, 88999 Kota Kinabalu, Sabah, Malaysia. d.bignell@qmul.ac.uk

Cares, J. E., Departamento de Fitopatologia, Instituto de Ciências Biológicas, Universidade de Brasília (UNB), Campus Universitário Darcy Ribeiro, Cx Postal 4457, CEP 70.910-900, Brasília, DF, Brazil. cares@unb.br

Cavalcanti, R. S., Departamento de Entomologia, Universidade Federal de Lavras, Cx Postal 3037, CEP 37200-000, Lavras, Brazil. rscavalcanti@yahoo.com.br

Coe, R., World Agroforestry Centre (ICRAF), United Nation Avenue, Gigiri, PO Box 30677-00100, Nairobi, Kenya. r.coe@cgiar.org

Constantino, R., Departamento de Zoologia, Universidade de Brasília (UnB), CEP 70910-900 Brasília, DF, Brazil. constant@unb.br

Csuzdi, C., Systematic Zoology Research Group of Hungarian Academy of Sciences and Hungarian Natural History Museum, H-1088 Budapest, Baross str. 13, Hungary. csuzdi-alef@nhmus.hu

Franklin, E., Coordenação de Pesquisas em Entomologia, Instituto Nacional de Pesquisas da Amazônia (National Institute for Amazonian Research-INPA), Cx Postal 478, CEP 69011-970, Manaus, AM, Brazil. beth@inpa.gov.br

Huang, S. P. (deceased), Departamento de Fitopatologia, Instituto de Ciências Biológicas, Universidade de Brasília (UnB), Cx Postal 4457, CEP 70.910-900, Brasília, DF, Brazil.

Huising, E. J., Tropical Soil Biology and Fertility (TSBF) Institute of CIAT, c/o ICRAF, United Nations Avenue, Gigiri, PO Box 30677-00100, Nairobi, Kenya. j.huising@cgiar.org

Karyanto, A., Faculty of Agriculture, University of Lampung, Jalan Sumantri Brojonegoro No 1, Bandar Lampung 35145, Indonesia. agsknila@yahoo.com

Konaté, S., Université d'Abobo-Adjamé, B.P. 801, Abidjan 02, Côte d'Ivoire. skonate2@yahoo.fr

Louzada, J. N. C., Departamento de Biologia (Biology Department), Universidade Federal de Lavras, Cx Postal 3037, Lavras, MG, Brazil, CEP 37200-000. jlouzada@gmail.com

Moino, A., Jr, Departamento de Entomologia, Universidade Federal de Lavras, Cx Postal 3037, CEP 37200-000, Lavras, Brazil. alcmoino@ufla.br

Morais, J. W., Coordenação de Pesquisas em Entomologia, Instituto Nacional de Pesquisas da Amazônia (National Institute for Amazonian Research-INPA), Cx Postal 478, 69011-970, Manaus, AM, Brazil. morais@inpa.gov.br

Moreira, F. M. S., Departamento de Ciência do Solo (Soil Science Department), Universidade Federal de Lavras, Cx Postal 3037, CEP37200-000, Lavras, MG, Brazil, CEP. fmoreira@ufla.br

Noordwijk, M. van, World Agroforestry Centre SE Asia, PO Box 161, Bogor, 16001 Indonesia. m.van-noordwijk@cgiar.com

Pfenning, L. H., Departamento de Fitopatologia, Universidade Federal de Lavras, Cx Postal 3037, CEP 37200-000 Lavras, MG, Brazil. ludwig@ufla.br

Rahmadi, C., Zoology Division, RC Biology, LIPI Jl., Raya Jakarta-Bogor Km. 46 Cibinong, Indonesia. cahyo.rahmadi@lipi.go.id

Silva, N. M., Faculdade de Ciências Agrárias, Universidade Federal do Amazonas, CEP 69070-000 Manaus, AM, Brazil. nmarques@ufam.edu.br

Stürmer, S. L., Departamento de Ciências Naturais, Universidade Regional de Blumenau, Cx Postal 1507, CEP 89010-971, Blumenau, SC, Brazil. sturmer@furb.br

Susilo, F.-X., Department of Plant Protection, Faculty of Agriculture, Universitas Lampung, Jalan Sumantri Brojonegoro No 1, Bandar Lampung 35145, Indonesia. fxsusilo@tlkom.net

Swift, M. J., Tropical Soil Biology and Fertility Institute of CIAT, c/o ICRAF, United Nations Avenue, Gigiri, PO Box 30677-00100, Nairobi, Kenya. swiftmj2003@yahoo.com

Tondoh, J. E., UFR des Sciences et de la Nature, Centre de Recherche en Ecologie, Université d'Abobo-Adjamé, 02 BP 801, Abidjan, Côte d'Ivoire. tondohj@yahoo.fr

Zanetti, R., Departamento de Entomologia (Entomology Department), Universidade Federal de Lavras, Cx Postal 37, CEP 37200-000, Lavras, MG, Brazil. zanetti@ufla.br

Foreword

Society has long known of its dependence on soils, but only recently has considered soils as having a biologically active component. By any measure, soils worldwide are in trouble, and there is a critical and immediate need to apply knowledge about the role of below-ground biodiversity in sustaining soils. The endorsement of the International Initiative for the Conservation and Sustainable Use of Soil Biodiversity by the Convention on Biological Diversity Conference of the Parties, in 2006, validated the importance of below-ground diversity in terrestrial ecosystem functioning and called for further multidisciplinary research. One focus arising from this must be how to assess and manage below-ground diversity in agricultural systems in order to facilitate agricultural productivity and long-term soil sustainability. This book unquestionably provides the best and most up-to-date effort to document how to assess soil biodiversity in ecosystems that are being rapidly altered by land use changes.

A major feature of global change in the tropical regions is land use change associated with agricultural intensification. Intensification is necessary to ensure global food supplies, but as intensification occurs the biological regulation of soil is altered and often substituted by fertilizers and mechanical tillage. This, in turn, frequently results in reduction of soil biodiversity and presents a challenge: is it possible to manage soil biodiversity to increase agricultural productivity in regions that are being degraded? The question is of critical importance for all people because land is being degraded not only in underdeveloped regions but in wealthy countries as well.

Ecologists have long debated the possible importance of soil biotic diversity, citing carbon sequestration in soils, reduction of greenhouse gas emissions, maintenance of soil physical structure and water retention capacity, nutrient provision to plants and control of plant pathogens as specific contributions of soil organisms to soil fertility. One square metre of temperate forest soil may contain as many as 1000 species of invertebrates, and perhaps a larger number of different types of micro-organisms. Tropical biodiversity is underestimated and true species densities may be even greater than at higher latitudes. The relationship between species diversity and functional diversity of the soil biota remains uncertain, but is the subject of investigation at all levels from the laboratory to the landscape. Likewise, the links between soil biodiversity and the occurrence and magnitude of ecological processes are uncertain. This knowledge is vital to the goal of achieving sustainable agricultural productivity with the conservation of existing biodiversity throughout the world, and not just in the tropics.

This book is intended to provide internationally accepted standard methods for the inventory of below-ground communities and characterization of land use in the humid tropics. It represents the published outcome of more than a decade of intense discussion and widespread fieldwork by a large number of expert scientists from many countries, and covers almost the full range of biological diversity from the largest invertebrates through to bacteria. The explicit aim is a practical one: to provide a definitive tool to establish a baseline and then to document the loss of soil biodiversity associated with deforestation and the process of agricultural intensification at forest margins. Ultimately these tools can be deployed to better use soil organisms in ecosystem management, in bioprospecting, and to achieve sustainable improvements in agricultural productivity for the world's poorest people. But the methods in this book extend to all terrestrial ecosystems as well.

Books of sampling methods for soil organisms have appeared at regular intervals for the past 50 years. This one has a number of important differences. Most importantly, the contributors are almost entirely drawn from countries that are most affected by deforestation and land use change; they represent the academic, institutional and governmental organizations where new policies for land management must be devised, validated and implemented if we are to have any chance of arresting the loss of tropical biodiversity. Inevitably, this will involve trade-offs between the concerns and interest of poor households, national development objectives and the quality of our environment. Secondly, the methods have all been developed and piloted under field conditions in 12 benchmark locations distributed across seven countries and three continents; consequently they are suitable not just for taxonomic experts but for general biologists, agriculture specialists and technical staff with all levels of training. As far as possible, the same applies to identification: the keys recommended are the ones that work best, and where it is advisable to refer specimens to a specialist, in country or abroad, this is explicitly recommended. Another notable feature of this book is that the soil biota are grouped into eight categories, each of which has a broad taxonomic identity but also corresponds to a major functional group considered important or even essential to soil function and therefore to the many ecosystem services that soils provide. The process has been necessarily selective, so some taxa have been excluded. Such selection is inevitable with soil, as the true biological diversity within it is staggeringly large and beyond the capacity of any scientific system to document fully, but has the advantage that it brings the means to make an approximate assessment of total soil health within the reach of those people who are most immediately concerned with its productivity. Finally, the authors have embraced the revolution in molecular methodology that has swept through the biological sciences during the last ten years. Many of the techniques associated with the assessment of diversity within specific microbial groups have emerged from the discussions amongst the eventual contributors to the book, and during workshops specifically arranged in preparation for this publication.

Diana H. Wall
Professor of Biology and Senior Research Scientist
Natural Resource Ecology Laboratory
Colorado State University
May 2008

Preface

This handbook is an outcome of the UNEP/GEF project 'Conservation and Sustainable Management of Below-ground Biodiversity' (CSM-BGBD) and emanated from the need for standard methods and practical instruction for the inventory of below-ground biodiversity (BGBD) that will allow the comparison of inventory results from benchmark areas in tropical countries on a valid scientific and statistical basis. The project was initiated in 2002 with the aim to generate information and knowledge that can be used to better manage and conserve BGBD in tropical agricultural landscapes in order to maintain agricultural productivity and reduce extension of agriculture into natural landscapes. The project is implemented in Brazil, Côte d'Ivoire, India, Indonesia, Kenya, Mexico and Uganda. It receives co-financing support from the Global Environment Facility (GEF) and implementation support from the United Nations Environment Programme (UNEP). The project is executed by the Tropical Soil Biology and Fertility Institute (TSBF); an institute of the International Centre for Tropical Agriculture (CIAT).

The handbook describes sampling and laboratory assessment methods for the biodiversity of a range of key functional groups of soil biota. It is a further elaboration and update of methods and protocols that were initially assembled and drafted by scientists affiliated with the Tropical Soil Biology and Fertility Institute of CIAT, the EU-funded Macrofauna Network, the NERC (UK)-funded Terrestrial Initiative in Global Environmental Research (TIGER), and in particular, the UNDP-GEF funded Alternatives to Slash-and-Burn project (ASB). Methods for some of the functional groups of soil organisms were included in a pioneering handbook of methods by Anderson and Ingram (1993). Methods for the assessment of soil biodiversity are elaborated further in the ASB Lecture Note 6B edited by Swift and Bignell (2001), that also introduces the principle of functional groups in the inventory of soil organisms. An authoritative manual of techniques for soil organisms and organisms inhabiting freshwater and marine sediments was published in 1996, edited by G. S. Hall. It was developed as part of UNESCO's contribution to the DIVERSITAS programme.

The methods described in this handbook reflect the wider scope of activities of the CSM-BGBD project in terms of the functional groups that are covered as well as broadly providing for the full geographical and biogeographical range of the humid tropics. It includes methods that were not included in the preceding publications mentioned above, such as those for mesofauna, entomopathogenic, saprophytic and pathogenic fungi and a wider range of the larger arthropods. It provides additional detail for specific functional groups, like the use of Winkler bags for ants and baited pitfalls for beetles in litter samples, or the updating of methods to evaluate diversity of nematodes, mycorrhizal fungi and nitrogen-fixing Leguminosae-nodulating bacteria (NFLNB) and their host plants. The proposed standard methods accommodate recent technological advances from molecular genetics, such as techniques for the genetic fingerprinting of NFLNB, where these are appropriate and within the reach of national laboratories.

Methods were discussed and progressively refined in annual meetings between 2002 and 2005, and their evolution is reflected in the reports of the annual project meetings of 2002 (Wageningen, The Netherlands), 2003 (Sumberjaya, Indonesia), and especially 2004 (Embu, Kenya), and in a number of taxon-specific or thematic workshops (Molecular techniques with emphasis on T-RFLP, Cali, Columbia, October 2003; Earthworms, Londrina, Brazil, May–June 2004 and Nairobi, Kenya, December 2004; AMF and Ectomycorrhiza, Bangalore, India, March 2005; Termites, Manaus, Brazil, August 2004; Termites and ants, Nairobi, February 2005) and in several in-country workshops held since the inception of the project. The definition of standard methods draws upon experience obtained with implementation of the methods in the seven countries participating in the project.

Because many species are still unknown or undescribed, the assessment of species diversity is not a very feasible undertaking. A more practical approach is to investigate the loss of key functional groups within the BGBD with change in land use or increasing land use intensity, to assess possible consequences of the loss in BGBD and interpret these processes in its ecological context. The standing hypothesis is that land use intensification results in a reduction of soil biodiversity leading to a loss of ecosystem services and is detrimental to sustained productivity. However, little is known about the precise relationships or about the mechanisms through which changes in BGBD occur. Land use conversion and intensification of land use are generally accompanied by drastic reductions in plant diversity, which we may expect to have implications for BGBD. At the same time, these land use conversions often result in a reduction of the soil organic matter that forms the basic substrate for most of the soil organisms, and will therefore influence both the abundance and diversity of these organisms. Apart from these effects, there are the direct effects of the management operations associated with increasing land use intensity. These are the disturbance of the soil as a habitat for the soil organisms from tillage and the direct effects of the enhanced use of agrochemicals. The effects on BGBD are compounded by environmental conditions, and soil physical and chemical characteristics (for example, pH, nutrient status and bulk density). Reversal of these trends or restoration of soil condition to its previous state will not automatically restore BGBD. Further, re-establishment of BGBD will not automatically restore the ecosystem services. Thus, many questions remain as far as management of BGBD is concerned. The premise is that agricultural diversification (at several scales) promotes soil biodiversity and that sustainable agricultural production (at tropical forest margins) is significantly improved by enhancement of soil biodiversity.

As by far the majority of the soil organisms are found in the upper 20cm of the soil profile, erosion may short-circuit all other degenerative processes and will have an immediate impact on BGBD that may easily overshadow the combined effect of all the above-mentioned factors and disturbances. Erosion of soil is therefore the ultimate and most catastrophic of disturbances, and erosion control and rehabilitation of badly eroded soils should be considered as part of strategies to conserve and sustainably manage BGBD.

Answering some of the above questions will require:

- Characterizing soil biodiversity occurring across a range of land uses of varying intensity, from natural forest to monoculture agriculture characterized by continuous cultivation and elevated levels of chemical input and mechanization.
- Establishing the relationship between the above-ground and the below-ground biodiversity across current and alternative land use systems and how this relationship is modified and altered by the prevailing environmental conditions (including soil conditions).
- Identifying 'entry points' for improved land management through the introduction and/or management of soil biota (experimenting with alternative land management options). The 'entry points' might include better understanding and use of indigenous practices, a wider knowledge of existing practices and more effective utilization of available technologies not requiring external inputs.

We hope the methods described in this handbook may assist in the design and execution of surveys for the inventory and monitoring of BGBD across scales from the plot (farmer's field, experimental or demonstration plot) to the landscape.

Fátima M. S. Moreira, E. Jeroen Huising and David E. Bignell

REFERENCES

Anderson, J. M. and Ingram, J. S. I. (eds) (1993) *Tropical Soil Biology and Fertility: A Handbook of Methods*, 2nd edition, CAB International, Wallingford

Hall, G. S. (ed) (1996) *Methods for the Examination of Organismal Diversity in Soils and Sediments*, CAB International, Wallingford

Swift, M. and Bignell, D. (eds) (2001) 'Standard methods for the assessment of soil biodiversity and land-use practice', International Centre for Research in Agroforestry, South East Asian Regional Research Programme, ASB Lecture Note 6B, Bogor, Indonesia, www.fao.org/ag/agl/agll/soilbiod/docs/manual-soil%20bioassessment.pdf

List of Acronyms and Abbreviations

ae	acid equivalent
ai	active ingredient
AFLP	amplified fragment length polymorphism
AMF	arbuscular mycorrhizal fungi
ANOVA	analysis of variance
ARA	acetylene reduction assay
ARISA	automated ribosomal intergenic spacer analysis
BF	bacterivores
CA	correspondence analysis
CFU	colony forming units
CMA	corn meal agar
cPCR	competitive polymerase chain reaction
CSM-BGBD	Conservation and Sustainable Management of Below-ground Biodiversity
D	disturbance
DGGE	denaturing gradient gel electrophoresis
DR	dilution rate
EBI	European Bioinformatics Institute
FF	fungivores
FG	functional group
FID	flame ionization detector
FSC	Forest Stewardship Council
FURB	Universidade Regional de Blumenau
GEF	Global Environment Facility
HV	high volume
IAR	intrinsic antibiotic resistance
ICRAF	World Agroforestry Centre
INPA	National Institute for Amazonian Research (Instituto Nacional de Pesquisas da Amazônia)
ITS	internal transcribed spacers
IU	infection unit
KOH	potassium hydroxide
LCCS	Land Cover Classification System
LV	low volume
MA	malt extract agar
MI	maturity index
MIP	mean infection percentage
MPN	most probable number

NCBI	National Center for Biotechnology Information
NFLNB	nitrogen-fixing Leguminosae-nodulating bacteria
PCA	principal component analysis
PCR	polymerase chain reaction
PDA	potato dextrose agar
PFGE	pulse field gel electrophoresis
PP	plant parasites
PPI	plant parasitic index
RAPD	random amplified polymorphic DNA
Rep-PCR	Repetitive Extragenic Palindromic elements/DNA sequences
RFLP	restriction fragment length polymorphism
RS	remote sensed (image)
SOM	soil organic matter
SRS	simple random sampling
SSCP	single strand conformation polymorphism
TGGE	thermal gradient gel electrophoresis
TRF	terminal restriction fragment
T-RFLP	terminal restriction fragment length polymorphism
TROF	trees on farm
TSBF-CIAT	Tropical Soil Biology and Fertility Institute of the International Center for Tropical Agriculture
UFAM	Universidade Federal do Amazonas
UFLA	Universidade Federal de Lavras
UFMG	Universidade Federal de Minas Gerais
UnB	Universidade de Brasilia
UNEP	United Nations Environment Programme

Chapter 1

The Inventory of Soil Biological Diversity: Concepts and General Guidelines

M. J. Swift, David E. Bignell, Fátima M. S. Moreira and E. Jeroen Huising

Soil organisms and soil ecosystem services

Soil is the habitat of an array of organisms in all three taxonomic domains (sensu Woese et al, 1990) and many phyla. The taxonomic classification of living organisms is still controversial (e.g. Margulis and Schwartz, 1998; Cavalier-Smith, 1998, 2004), especially regarding the taxa to be created at higher levels, and the numbers of such higher categories (such as kingdom) to be considered, in addition to domain. However, whatever classification system is used, the diversity of soil biota is high at all levels of analysis (for reviews, see Swift et al, 1979; Lavelle, 1996; Brussaard et al, 1997; Wall, 2004, Bardgett, 2005; Moreira et al, 2006). Table 1.1 lists the main phyla of eukaryotic organisms and prokaryotes that are or can be represented in the soil community, with more than 1.5 million species for the eukaryotes and an estimated species richness far beyond 10,000 for the prokaryotes.

As it is neither practicable nor sensible to address all the organisms present when making an assessment of the biological health of soils (see Lawton et al, 1998), biota have been evaluated (relative to non-biotic agencies and between themselves) for their relative contribution to ecosystem processes (sensu Daily, 1997; Wall, 2004). These processes collectively support the provision of ecosystem services that contribute to the maintenance and productivity of ecosystems by their influence on soil quality and health (Hole, 1981; Lavelle, 1996, Brussaard et al, 1997; Kibblewhite et al, 2008). The processes can be grouped into four main aggregate ecosystem functions:

1 *Decomposition of organic matter* is largely brought about by the enzymatic activity of bacteria and fungi, but greatly facilitated by soil animals such as mites, millipedes, earthworms and termites, which shred the plant or animal residues and disperse microbial propagules. Together, the microorganisms and the animals involved in the

Table 1.1 *Number of described species in the main taxonomic categories of plants and of soil biota, and the main functional groups they represent*

Taxonomic categories[a] *[Total number of extant phyla] (examples of soil organisms/common name)*	*No of described species in the taxon (all habitats)*	*Functional group*[b]
Domain Eucarya		
Kingdom Plantae [12 phyla]	255,000	
Phylum Bryophyta (mosses)	10,000	1
Phylum Hepatophyta (liverworts)	6000	1
Phylum Filicinophyta (ferns)	12,000	1
Phylum Cycadophyta (gymnosperms)	185	1
Phylum Coniferophyta (gymnosperms)	550	1
Phylum Gnetophyta	70	1
Phylum Anthophyta (angiosperms)	235,000	1
Monocotyledons	65,000	1
Dicotyledons	130,000	1
Kingdom Animalia [37 phyla]	>10 million	
Phylum Tardigrada[d] (water bears)	750	2, 4
Phylum Mollusca[d]	99,000	
Class Gastropoda[d] (includes slugs and snails)	35,000–40,000	2, 4
Phylum Annelida[d] (segmented worms)	>18,000	
Class Polychaeta	9000	Very rare in soil
Class Oligochaeta (earthworms, enchytraeids)	8800	3, 4
Class Hirudinea (leeches)	500	4
Phylum Crustacea[d] [>6 classes]	45,000	
Class Malacostraca[d] (includes Order Decapoda with a hard calcified external skeleton)	25,000	Common only locally
Order Isopoda (wood mites, woodlice)	>11,000	5
Phylum Mandibulata (Arthropoda)		
Sub-Phylum Hexapoda (Insecta)	>900,000	
Order Archaeognata (bristletails)	350	4
Order Thysanura	700	4, 6
Order Blattoptera (cockroaches)	4000	4
Order Dermaptera (earwigs)	1800	2, 4, 6
Order Hemiptera	>80,000	2, 9
Sub-Order Homoptera (planthoppers, cicadas, aphids, shield bugs, true bugs)	55,000	2, 9
Sub-Order Heteroptera	25,000	2, 9
Order Isoptera (termites)	2800	3, 4
Order Orthoptera (grasshoppers, crickets, locusts)	23,000	2, 9
Order Thysanoptera (thrips)	6000	2, 9
Order Coleoptera (beetles)	>350,000	2, 4, 6, 9
Order Diptera (flies, mosquitos, gnats, midges)	>125,000	2, 4, 6, 9
Order Hymenoptera (ants, bees, wasps, sawflies)	115,000	
Family Formicidae (ants)	11,826	2, 3, 6
Order Lepidoptera (butterflies, moths, skippers)	180,000	2, 9
Order Trichoptera	12,000	Riparian only
Order Collembola (springtails)	7500	4, 7
Order Diplura (two-pronged bristletails)	659	2, 6
Order Protura	500	4, 7

Sub-Phylum Myriapoda (millipedes and centipedes)	15,162	
Class Diplopoda (millipedes)	10,000	4, 9
Class Chilopoda (centipedes)	2500	6
Class Symphyla	200	2, 7, 9
Class Pauropoda	700	4, 7
Sub-Phylum Chelicerata[d] [3 classes]	>100,000	
Class Arachnida [11 orders]	93,455	
Order Palpigradi (micro-whipscorpions)	80	6
Order Acari (mites)	45,000	2, 4, 6, 7, 9
Order Pseudoscorpionida (pseudoscorpions)	3235	6
Order Araneae (spiders)	40,000	6
Order Scorpionida (scorpions)	2000	6
Phylum Gastrotricha[d] (gastrotrichs)	400	7
Phylum Acanthocephala[d] (spiny-headed worms)	1000	Arthropod parasite
Phylum Rotifera[d] (wheel animals)	2000	4, 7
Phylum Nemertina[d] (ribbon worms)	900	Very rare in soil
Phylum Nematoda (nematodes, roundworms, pinworms)	15,000	2, 6, 7, 9
Phylum Plathyhelminthes[d] (helminthes, includes turbellarians)	25,000	2, 6
Kingdom Protoctista[c] [30 phyla]	Undetermined number	
Phylum Rhizopoda (amoebae, slime-moulds)	Undetermined number	7
Phylum Dinomastigota[d] (dinoflagellates)	4000	1, 7
Phylum Ciliophora (ciliates)	10,000	7
Phylum Discomitochondria (flagellates and zooflagellates)	800	1, 7
Phylum Diatomacea[d] (diatoms)	10,000	1
Phylum Oomycota (oomycetes)	Hundreds	5, 9
Phylum Rhodophyta (red algae)	4100	1
Phylum Chlorophyta (green algae)	16,000	1
Kingdom Fungi [6 phyla]	>70,000	
Phylum Microsporidia	1500	Parasites only
Phylum Chytridiomycota (chytrids)	1000	5, 9
Phylum Zygomycota (moulds)	1100	5
Phylum Glomeromycota (arbuscular mycorrhizal fungi)	204	5, 8
Phylum Basidiomycota (includes mushrooms and some yeasts)	>22,250	5, 8
Phylum Ascomycota (includes yeasts)	>30,000	5, 8
Domain Archaea[d] [4 phyla]	>344[f]	10
Domain Bacteria[e] [52 phyla]	>8398[f]	1, 5, 8, 9, 10

Notes: a Considering taxonomic categories from the highest to the lowest level: domain, kingdom, phylum, class, order, family, genus, species. Prokaryote (Domains Archaea and Bacteria) classification according to Woese et al (1990). Eucarya kingdoms classified according to Margulis and Schwartz, 1998. Fungi according to James et al (2006).

b Functional groups according to Table 1.2 (this chapter).

c Considered by some authors as Protists or Protozoa and Chromista.

d Includes soil and aquatic organisms.

e Rappé and Giovannoni (2003).

f National Center for Biotechnology Information: (www.ncbi.nlm.nih.gov, accessed 13 May 2007), excluding unclassified, uncultured and unspecified organisms; if these are included the corresponding richness will be 3090 and 79,342, respectively.

Source: After Moreira et al, 2006

process are called decomposers, but the term litter transformers has now come to be used to describe these animals, where they are not also ecosystem engineers (see below). As a result of decomposition, organic C is released into the atmosphere, predominantly as CO_2 or CH_4, but also incorporated into a number of pools within the soil as soil organic matter (SOM). These SOM fractions vary in their stability and longevity, but within a given soil type and environment a characteristic equilibrium exists between the SOM content and the inflows and outflows of C from the system.

2 *Nutrient cycling*, which is closely associated with organic decomposition. Here again the microorganisms mediate most of the transformations, but the rate at which the process operates is determined by small grazers (micropredators) such as protoctists, nematodes, collembolans and mites. Larger animals may enhance some processes by providing niches for microbial growth within their guts or excrement. Specific soil microorganisms also enhance the amount and efficiency of nutrient acquisition by the vegetation through the formation of symbiotic associations such as those of mycorrhiza and N_2-fixing root nodules. Nutrient cycling by the soil biota is essential for all forms of agriculture and forestry. Some groups of soil bacteria are involved in autotrophic elemental transformations, that is, they do not depend on organic matter directly as a food source, but may nonetheless be affected indirectly by such factors as water content, soil stability, porosity and C content, which the other biota control.

3 *Bioturbation*. Plant roots, earthworms, termites, ants and some other soil macrofauna are physically active in the soil forming channels, pores, aggregates and mounds, and moving particles from one horizon to another. These processes of 'bioturbation' influence and determine soil physical structure and the distribution of organic material. In so doing they also create or modify microhabitats for other, smaller, soil organisms and determine soil properties such as aeration, drainage, aggregate stability and water holding capacity. This set of organisms has therefore been called 'soil ecosystem engineers' (Stork and Eggleton, 1992; Jones et al, 1994; Lawton, 1996; Lavelle et al, 1997). Soil structure and properties are also influenced though the production by the animal engineers of faeces, comprising organo-mineral complexes that are stable over periods of months or more (Lavelle et al, 1997). Bioturbation plays a major role in the regulation of the water balance of the soil (infiltration, water storage capacity and drainage) and strongly influences its susceptibility to erosion.

4 *Disease and pest control*. The soil biota includes a wide range of viruses, bacteria, fungi and invertebrate animals capable of invading plants and animals (including humans) and causing disease and death. In natural ecosystems, intensive outbreaks of soil-borne diseases and pests are relatively rare, whereas such epidemics are common in agriculture. In healthy soils the activities of the potential pests and pathogens are regulated by interactions with other members of the soil biota, which include microbivores and micropredators that feed on microbial and animal pests respectively, as well as a wide variety of microbial antagonistic interactions. In agroecosystems this range of interactions may be reduced because of diminished biological diversity and/or soil environmental changes such as those caused by lowered SOM content.

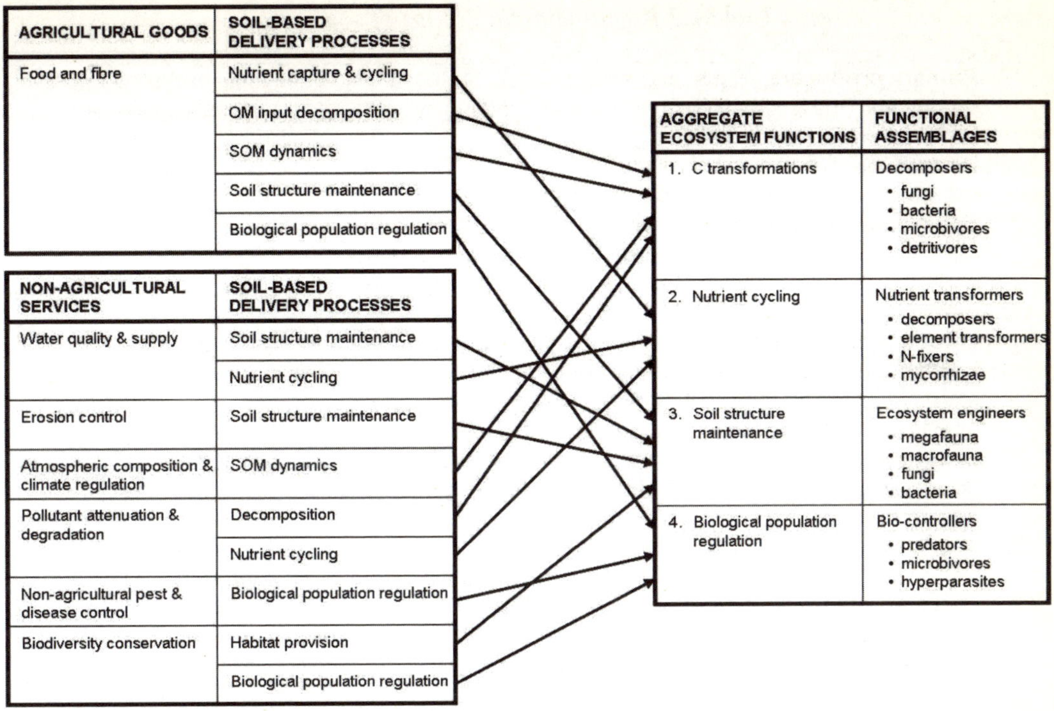

Source: From Kibblewhite et al, 2008

Figure 1.1 *The relationship between the activities of the soil biological community and a range of ecosystem goods and services that society might expect from agricultural soils*

Figure 1.1 shows the contribution made by the soil biota to ecosystem goods and services as a result of the above processes. In particular it should be noted that the interaction between organic matter decomposition, bioturbation and nutrient cycling will determine the balance between the equilibrium amount of carbon sequestered in the soil (see above) and the emissions of greenhouse gases (principally CO_2, CH_4, NO_x, N_2O). Soil organisms thus play an important role in the regulation of atmospheric composition and hence are major players in climate change.

Functional groups of soil biota

In principle, all the organisms listed as members of the soil community can be allocated to one or more of the four generic functional categories described above, based on the particular function they perform or the specific soil-based process they mediate. In order to link particular soil organisms (collectively soil biodiversity) to the generic functional categories and ultimately the ecosystem services (e.g. Setäla et al, 1998), we resort to the concept of key functional groups, usually defined on trophic criteria (Brussaard, 1998) but qualified by morphological, physiological, behavioural, biochemical or environmental responses, and to some extent by taxonomic character.

Table 1.2 *Key functional groups of soil biota*

1. **Primary producers** (higher and lower plants): photosynthetic organisms assimilating carbon dioxide from the air, penetrating the soil with rooting systems and translocating organic compounds synthesized above ground.
2. **Herbivores:** animals consuming and partly digesting living plant tissues, including leaf miners and rollers, stem borers and sap suckers.
3. **Ecosystem engineers** (e.g. macrofauna such as termites and earthworms): organisms that have major physical impact on soil through soil transport, building of aggregate structures and formation of pores – as well as influencing nutrient cycling. Can include predators (e.g. many ants).
4. **Litter transformers** (many macrofauna and mesofauna, but some microfauna): invertebrates feeding on microbially conditioned organic detritus and shredding this material (comminution) and making it more accessible to decomposers, or promoting microbial growth in pelletized faeces. This activity can be performed at several spatial scales.
5. **Decomposers** (e.g. cellulose-degrading fungi or bacteria): micro-organisms possessing the polymer-degrading enzymes that are responsible for most of the energy flow in the decomposer food web.
6. **Predators** (many macrofauna and mesofauna): animals which regulate herbivores, ecosystem engineers, litter transformers, decomposers and microregulators through predation.
7. **Microregulators** (e.g. microfauna such as nematodes): animals that regulate nutrient cycles through grazing and other interactions with the decomposer microorganisms.
8. **Microsymbionts** (e.g. mycorrhizal fungi, rhizobia): microorganisms associated with roots that enhance nutrient uptake.
9. **Soil-borne pests and diseases** (e.g. fungal pathogens, invertebrate pests): biological control species (e.g. predators, parasitoids and hyperparasites of pests and diseases) can also be included.
10. **Prokaryotic transformers:** archaea and bacteria performing specific transformations of carbon (e.g.methanotrophy) or nutrient elements such as N, S or P (e.g. nitrification, nitrogen fixation).

There is no precise agreement on the definition of the functional groups and how many of these groups should be discerned within in a typical soil environment, so the concept is heuristic and can be modified for whatever analytical purpose is in hand, but a good case can be made for at least ten categories. These are presented formally in Table 1.2 and used to annotate the taxonomic categories listed in Table 1.1. Note that some functional groups include a wide range of related and sometimes unrelated taxa, while others have high taxonomic specificity. For this reason, and because many components of soil biota are taxonomically intractable, there have been few studies of agricultural soil health in which the whole taxonomic spectrum has been sampled representatively in the same place and at the same time (Bignell et al, 2005).

TARGET GROUPS OF SOIL BIOTA

In designing fieldwork the challenge is to select a subset of the soil biota that adequately reflects the anticipated taxonomic spectrum, and which at the same time includes all the functional groups considered important. The functional importance of any species or group of species is likely to be related to their relative abundance and biomass, so assess-

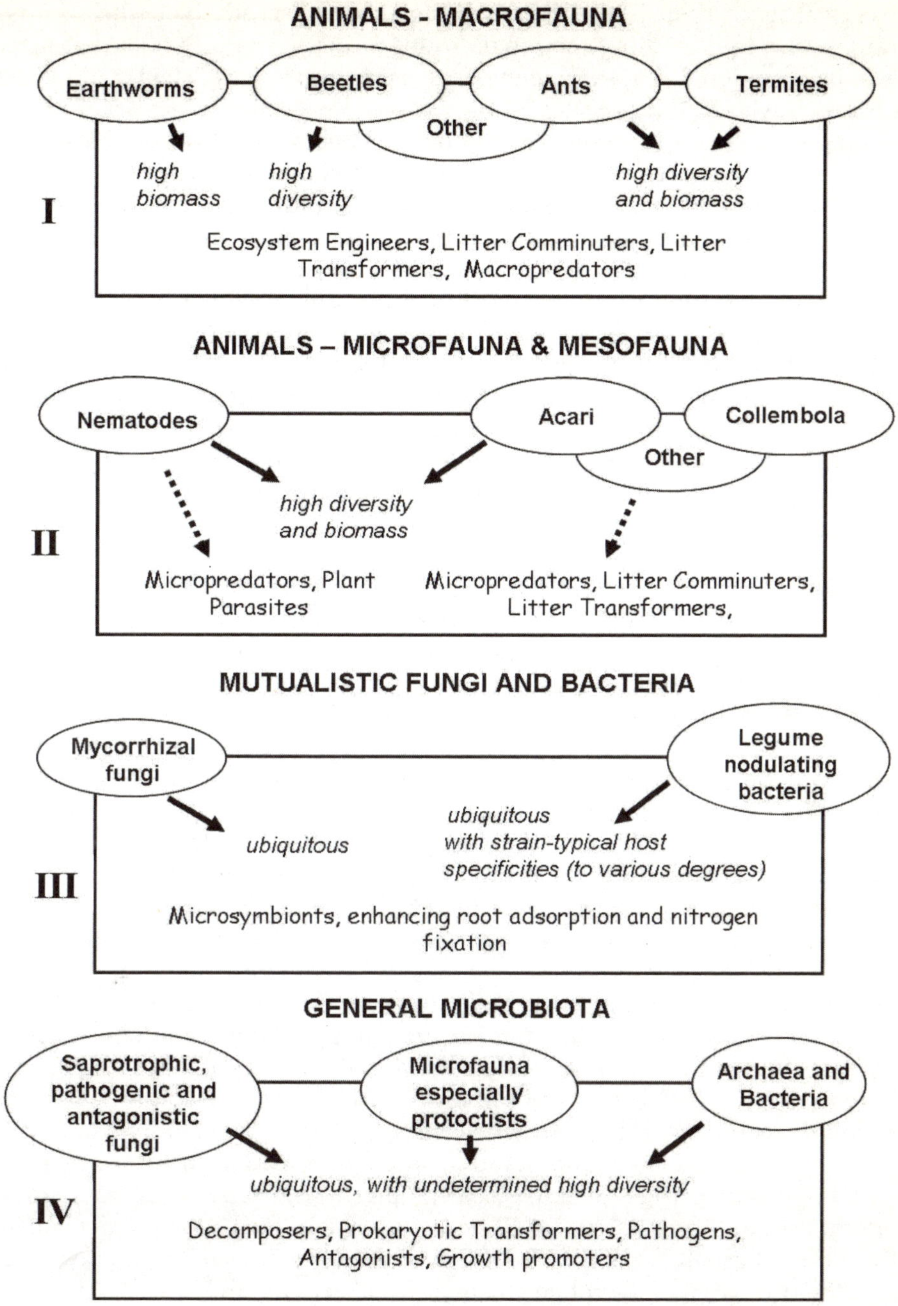

Note: Fauna are classified according to body width: macrofauna >2.0mm; mesofauna 0.1–2.0mm; microfauna <0.1mm. Alternatively, nematodes and protoctists can be considered as components of the general microbiota, a group which includes many organisms with specialist roles, as well as a diversity of generic decomposers.

Figure 1.2 *Main functional groups studied in the CSM-BGBD project classified according to domains and kingdoms, size and related ecosystem processes*

ing presence/absence alone is insufficient to the above purposes. Nonetheless it is also important to also look within functional groups to discover those taxa that meet the criteria to be considered ecosystem engineers (sensu Jones et al, 1994), and keystone species (sensu Davic, 2003).

The taxonomic groups nominated below are thus selected on the basis of their diverse functional significances to soil fertility and soil quality (hence the term 'target taxa'); and their relative ease of sampling, isolation and identification (Figure 1.2). These are the taxonomic groups that were addressed in the Conservation and Sustainable Management of Below-ground Biodiversity project (CSM-BGBD) and for which methods of inventory and characterization are provided in the subsequent chapters. Furthermore, a number of these taxonomic groups are very important in their own right as contributors to overall biotic diversity, for example beetles, ants, enchytraeid worms, spiders, mites, nematodes and (probably) protoctists (protists), all of which comprise large numbers of species in comparison with other higher taxonomic groups (Table 1.1). In this context, the geographical distribution, the introduction of exotic and invasive species and loss of endemic species then also become a concern.

The target groups included are:

1 *Macrofauna – earthworms*, which influence both soil porosity and nutrient relations through tunnelling, and ingestion of mineral and organic matter, and which act as regulators of soil biotic populations at smaller spatial scales, for example mesofauna, microfauna and microsymbionts.
2 *Macrofauna – termites*, *ants* and *beetles*, which influence or mediate a) soil porosity and texture through tunnelling, soil ingestion and transport, and gallery construction, b) nutrient cycles through transport, shredding and digestion of organic matter, and c) biological control as predators.
3 *Other macrofauna* such as woodlice (Isopoda), millipedes (Diplopoda) and some types of insect larvae, which act as litter transformers, with an important shredding action on dead plant tissue, and their predators (centipedes, larger arachnids, some other types of insect). Some species can also have detrimental effects by becoming pests.
4 *Mesofauna*, such as collembolans and mites, which act as litter transformers and micropredators (grazers of fungi and bacteria, and predators of other soil animals), thus contributing to smaller-scale organic comminution processes and exerting a strong regulatory role within the soil biota. Some species of mites and the larval forms of many soil arthropods are small enough to be classified as microfauna.
5 *Microfauna* – such as nematodes and protoctists, which a) influence turnover in their roles as root grazers (plant parasites), fungivores, bacterivores, omnivores and predators, b) occupy existing small pore spaces in which they are dependent on water films, c) usually have very high generic and species richness, d) have a strong role in the regulation of microbial abundance and activity, and e) as insect pathogens are important as a source of biological control.
6 *Arbuscular mycorrhizal fungi*, which associate with plant roots, improving nutrient availability, and reducing attacks by plant pathogens as well as improving tolerance to environmental stresses.
7 *Legume-nodulating bacteria* and, when relevant, other N-fixing microsymbionts, which transform atmospheric N_2 into NH_3, which is a form available for plant growth.

8 *Phytopathogenic, saprotrophic* and *antagonistic fungi*, which determine or mediate (in different cases) crop viability, the turnover of organic carbon during decomposition, and plant diseases, and contribute to the potential biological control of pests and diseases.

Methods for inventory of below-ground biodiversity: General principles

Current methodology does not allow for the identification of all species in the soil environment. In many phyla the majority of species are still unknown. The species concept is also highly variable across the soil biota – for instance between the bacteria, the fungi and the invertebrate animals. In the cases of the Archaea and Bacteria, molecular and biochemical characterization now replaces traditional identification approaches. The methods employed for extracting the different groups of organisms from the soil are also very different from one group to another. These factors, plus the sheer scale of the diversity found in the soil, necessitate a selective approach to below-ground biodiversity inventory, using the concept of key functional groups as described above.

It is as yet an unresolved question what relationship exists between species diversity, functional diversity (the number of functional groups), functional composition (the nature of the functional groups) and the occurrence and intensity of ecological processes. One would expect ecosystem services that rely on functional groups that are composed of relatively few species and/or are highly specialized to be most vulnerable to stress and disturbances that affect these functional groups. To the best of our knowledge this holds for shredders of organic matter (see litter transformers in Table 1.2), nitrifying and denitrifying bacteria (as part of the prokaryotic transformers), bacteria involved with C1 compound and hydrogen transformations (e.g. methanogenic, methylothrophic), iron, copper and sulphur chemolithotrophs, mycorrhizal fungi (among the microsymbionts) and bioturbators (i.e. ecosystem engineers) – in contrast with the high diversity and functional redundancy of the majority of decomposers. This invokes questions like 'what is the minimum number of species within functional groups to ensure soil ecosystem resilience against natural and anthropogenic stress', and in relation to that 'what are the key species (if any) within functional groups'. This invites evaluation of the effectiveness of an individual species in performing a particular function. Therefore, when adopting the concept of key functional groups, questions like those above should be considered in defining the purpose for the investigation of below-ground biodiversity, and will reflect upon the methods to be used.

Standard methods are needed to describe ecosystems in a consistent manner. This manual addresses the methods for the inventory and characterization of below-ground biodiversity, focusing on functional groups and allowing for assessment of diversity, abundance and species composition. Assessments of biotic community compositions are dependent on the positive identification of the collected specimens; this requires a high level of available taxonomic expertise, which for some groups and in some countries is a major challenge. Taxonomic issues are not addressed in this handbook, though references are provided for keys to the identification of species in the various functional groups.

More specifically, the methods have been developed or chosen and modified for use in landscapes of varying land use in order to provide means of assessing the impact of (for instance) agricultural intensification, habitat fragmentation and cropping management on soil biological diversity, but should serve to answer a range of questions in similar contexts. A brief description of the project for which the methods were developed and tested is given in the Preface.

INVENTORY OF BELOW-GROUND BIODIVERSITY ACROSS SPACE AND TIMESCALES

Land use change as the primary driver of changes in soil biological diversity

An inventory may demonstrate whether soil biota are responsive to human-induced disturbances such as agricultural practices, deforestation, pollution and climate change. If so, there may then be many negative consequences in terms of diminished delivery of ecosystem services, including loss of primary productivity through changes in soil fertility and/or increases in soil-borne disease, loss of cleansing potential for wastes and pollutants, disruption of global elemental cycles, and feedbacks on greenhouse gas fluxes and erosion. Global food supply depends on intensive agriculture; as intensification proceeds above-ground biodiversity is reduced, with consequences for the below-ground biodiversity and thence the biological regulation of soil-based ecosystem services. These regulatory functions are often described as being 'substituted' by inputs such as the use of mechanical tillage, chemical fertilizers and pesticides (Kibblewhite et al, 2008). This is also assumed to reduce below-ground diversity, which if accompanied by the extinction of individual species may cause losses of function and reduce the ability of agricultural systems to withstand unexpected periods of stress, bringing about undesirable effects. Large numbers of farmers in the tropics have limited access to inputs but in an attempt to intensify production the complexity of their agroecosystems has become drastically reduced nonetheless. An alternative solution is to intensify while at the same time retaining a greater degree of above-ground diversity. The maintenance of diversity of crops and other plants in cropping systems is widely accepted as a management practice that buffers farmers against short-term risk. Enhanced biodiversity and complexity above ground contributes to the re-establishment or protection of the multiplicity of below-ground organisms able to carry out essential biological functions. This can be considered at both the field and the landscape level to enhance structural complexity and functional diversity, especially in degraded lands.

What to sample for, where and when

Although this is a manual of practical instruction, such practice must as far as possible reflect the consensus of ecological theory, and in particular current thinking about community structures at different spatial and temporal scales that determine the provision of ecosystem services. However, the processes of disturbance (and recovery) that affect below-ground biodiversity also manifest themselves at different spatial and temporal scales (i.e. various frequencies) that are often linked. The central tenet is that the functional significance of biological diversity changes across scales in space and time (Swift et al, 2002, Lavelle et al, 2004). Broadly, this means that what is measured, and where and when it is measured, has to take account of the features of biodiversity that are likely to affect overall ecosystem services in different types of comparison.

The aim should be to reduce intra-group variability and maximize variability between groups. The major challenge then is how to define these 'groups' in relation to spatial and temporal scale levels such that this is achievable. For example, we may expect a large variation in numbers and diversity of earthworms, moving from one plot to the other or from the field to the hedge at field boundary, while for the land use zone, that combines these elements within the land use system, the numbers or diversity may remain stable. That is to say that inventory of the earthworm species of any particular plot may fall well short of the purpose of assessment of the diversity of the whole area, as might be needed if the objective of the inventory is for conservation of below-ground biodiversity. Likewise, the difference in decomposition rate or symbiotic nitrogen fixation between two spadefuls of soil from the same field is probably a question of which microorganisms are present in each, but between two adjacent fields in the same landscape it could be a matter of drainage or site management history or equally well whether termites or some other ecosystem engineer were more active in one site than the other.

Between two landscapes or two regions we would expect climate and parent material to be the main determining factors in soil forming processes. Soil formation at the regional scale is one of the ecosystem services that integrate processes over all scales. The climate, parent materials and soils will be the major determinants of all ecosystem services at the landscape level, but there might be a significant biotic influence if one or more important functional group were missing or depleted in a particular one location, or if the parent forest ecosystem had been reduced to a relict in one place but maintained in a mixed landscape at the other.

Temporal scales are important for degradation processes (for example the progressive loss of macropores, the erosion of finer soil particles and the depletion of complex organic matter may ultimately destroy productive capacity beyond recovery) and also for rehabilitation (the life cycles of different biotic taxa vary from hours to years, and may incorporate obligate dormancies, while fecundities also vary and therefore the time required for a growing population to reach a minimum critical size where it can become self-sustaining will also be different from one organism to another). The main theoretical concepts that correspond to these concerns are openness and connectivity, and are discussed in detail for soil and sediment ecosystems by Wall (2004). Their impacts on the design of our own sampling and measurement protocols are strongly reflected in Chapter 2 of this manual.

The four levels of spatial scale for sampling given below are those used in the CSM-BGBD project (see Preface) and are given as an example. Every project will need to determine the scales appropriate to its own objectives. Table 1.3 provides a summary of the hierarchy levels (modified from those proposed by Lavelle et al, 2006), the related processes of change that will effect below-ground biodiversity (and associated ecosystem services) and the component of below-ground biodiversity of particular relevance at that level.

1 *Landscape.* At the highest level the spatial distribution of below-ground biodiversity will be strongly influenced by gradients in altitude (and temperature), gradients or transitions in moisture and gradients in soil properties derived from mineral base and vegetation. The land use pattern superimposed on these effects of climate,

geology, vegetation and soil will be a further major determinant at this level. The spatial characteristics of the land use distribution (e.g. patchiness of the land use components like forest) will also have an important influence on the biodiversity. The inventory of below-ground biodiversity at a landscape scale must therefore capture the distribution, sizes and shapes of these gradients and patches in order to explain possible variation in below-ground biodiversity.

At the landscape level it is difficult to evaluate the effect of below-ground biodiversity on the provision of some ecosystem services like 'water quality and supply' or 'erosion control' and it is unlikely to be important compared with effects of the vegetation. Below-ground biodiversity, including the discovery of exotic invading species, is in this case thus more useful as an indicator of soil response to management. Protection and conservation of key (agro) ecosystems and habitats are the main option for intervention. Timescales for these processes (e.g. change in land use composition and landscape configuration) to take effect may vary between years (exceptionally) and hundreds of years.

2 *Land use zone*. Land use zones are defined as areas with a particular land use and management regime (Huising, 1993), for example a zone with commercial tea plantations or an area characterized by smallholder subsistence agriculture. These zones are typically characterized by a particular combination of land-cover types. The above-ground biodiversity is determined by the variety of crops grown in the area and vegetative elements associated with fallows, windbreaks, hedgerows, woodlots and other managed landscape features. Lateral interactions are an important component for below-ground biodiversity and the way the system functions, most notably when the agriculture is agroforestry. Gradients in soil properties do occur at this level, but are associated with management; for example Vanlauwe et al (2006) and Tittonell et al (2007) describe soil fertility gradients as the result of variable resource allocation within farm. It is, therefore, at this level that below-ground biodiversity should be studied in the greatest detail with respect to land use change. Land use intensification is typically associated with the removal of vegetative elements such as on-farm trees, with loss of their inputs and impacts, for instance the niches for particular soil organisms which the trees provide. Thus the relation between the spatial distribution patterns of the above-ground biota and those of below-ground species are of particular interest. At this level the timescales in which changes related to land use and land cover characteristics occur are shorter, ranging from a few years to several decades.

3 *Plot or patch*. At the plot level the functional composition and activities of the biota are of particular interest and, as far as the inventory of below-ground biodiversity is concerned, so are the relative abundances of species belonging to the various trophic levels and functional groups that govern the stability of foodwebs and provision of support functions. At this level, community structure and interactions are regulated by the local (plot-specific) architecture of the soil (Lavelle et al, 2006). The plot or patch (if not too big) is within the functional domains of ecosystem engineers, including plant roots. Changes in the soil architecture, whether brought about by changes in the functional composition and activity of the ecosystem engineers or tillage of the soil, will have an effect on the trophic interactions and thereby on the (structure of the) foodweb. Organic matter input management and the use of agrochemicals will further impact the foodweb structure.

Inoculation of any kind directly impacts below-ground biodiversity and can therefore be aimed to correct for imbalances in the functional compositions (or even to enhance particular functions) and the foodweb (e.g. introduction of predators to control a particular disease). Such intervention should therefore be informed by the inventory and relative abundances of species belonging to defined functional groups or levels in the food chain. At the plot level the underlying delivery processes mediated by soil organisms for the provision of support services, such as soil structure maintenance and organic matter input decomposition, have relatively short timescales and will show seasonal variation. Structural changes within the biotic community (foodweb connections, functional composition) may take place within years.

4 *Sub-plot niche or microhabitat.* This level refers to the microhabitats that accommodate the microbes and the microfaunal components of the soil system which regulate them. These exist at the scale of the soil horizon, the individual aggregate and the individual particle, and are segregated into functional domains (e.g. rhizosphere, drilosphere, termitosphere, etc.), reflecting the dependence of the smallest organisms on the ecosystem engineers for the provision of their substrates (for detailed concepts and terminology see Lavelle and Spain, 2001). Functional redundancy (different species or strains performing the same function) amongst the primary transformers, competition and efficiency are of particular relevance here, and in relation to this the degree of genetic diversity within the microorganisms is also of interest. Options for management interventions at the sub-plot scale are currently limited, mainly inoculation with microbial species, but an understanding of the processes and interactions is ultimately vital, as this is the most fundamental unit of soil function.

The 'interventions' listed in Table 1.3 are partly derived from the concept of 'hierarchical management of soil biota' as described by Swift (1998). The list of interventions and measurements related to below-ground biodiversity does not aim to be exhaustive, but rather illustrative, and there is a certain overlap between scale levels in the parameters to be measured.

Data requirement

The methods in this manual provide for the assessment of a large number of parameters of below-ground biodiversity. Irrespective of the specific objectives and experimental design of any study, data will still have to be gathered per sampling point, for which the best available methods consistent with a sensible use of time and resources are given in Chapters 3 to 10. Overall, the inventory of below-ground biodiversity using the methods in this manual may provide the following data and information, depending on the type of analyses performed:

Concerning biodiversity:

- taxonomic richness at species or higher, or lower (e.g. strain for microsymbiont) level;
- abundance (all) and biomass (for animals) of taxa;
- abundance and biomass of functional groups (FGs);

Table 1.3 *Hierarchical levels in inventory and management of below-ground biodiversity, reference units, associated process of change and relevant biodiversity components*

Layer or aggregation level	*Type of unit*	*Number of classes or units*	*Process/intervention*	*Measurement/ biodiversity parameter*
Landscape	Area characterized by mosaic of land uses and of ecosystems like larger watershed	Unspecified; maybe several per biome or region within a country	Land use conversion; protection of land uses and landscape elements of particular relevance to biodiversity and to maintain provision of ecosystem services (e.g. forest patches, biological corridors)	Soil biological diversity; exotic and invasive species
Land use zone	Area with characteristic land use and land cover pattern and management regime	3–20 types of land use zones within a particular landscape (depending on the level of detail of the classification)	Land use intensification, land degradation; maintaining biological refuge areas and above-ground biodiversity, Integrated Natural Resources Management and land conservation practices	Functional diversity, species diversity and composition esp. with respect to ecosystem engineers; spatial distribution pattern
Plot, Patch	Area with a particular use and management, e.g. agricultural field, vegetation plots, woodlot, etc.	10s to 100s per land use zone	Intensification of land and crop management, erosion; improved cropping systems and organic matter management, minimum tillage etc., inoculation	Relative abundance of species within and between functional groups and trophic levels; functional composition
Niche, Microhabitat	e.g. soil horizons, rhizosphere, drilosphere		Management of keystone biota; introduction of biological control agents, inoculation with microsymbionts	(Microbial) biomass and activity; genetic diversity

- relative proportions of FGs;
- overall quantitative indices of richness and evenness.

Concerning soil, land use, land use intensity, and land cover:

- apparent cropping or fallow usage;
- current land and crop management practices;
- basic physical and chemical soil properties; slope and aspect;

- above-ground vegetation character;
- climatic averages and actual rainfall to sampling date;
- precise history of use since undisturbed forest or other pre-cultural vegetation.

References

Bardgett, R. D. (2005) *The Biology of Soil*, Oxford University Press, Oxford

Bignell, D. E., Tondoh, J., Dibog, L., Huang, S. P., Moreira, F., Nwaga, D., Pashanasi, B., Susilo, F. X. and Swift, M. (2005) 'Belowground diversity assessment: Developing a key functional group approach in best-bet alternatives to Slash-and-Burn', in C. A. Palm, S. A. Vosti, P. A. Sanchez, P. J. Ericksen (eds) *Slash-and-Burn Agriculture: The Search for Alternatives*, Columbia University Press, New York

Brussaard, L. (1998) 'Soil fauna, guilds, functional groups and ecosystem processes', *Applied Soil Ecology*, vol 9, pp123–135

Brussaard, L., Behan-Pelletier, V., Bignell, D., Brown, V., Didden, W., Folgarait, P., Fragoso, C., Freckman, D., Gupta, V., Hattori, T., Hawksworth, D., Klopatek, C., Lavelle, P., Malloch, D., Rusek, J., Soderstrøm, B., Tiedje, J., and Virginia, R. (1997) 'Biodiversity and ecosystem functioning in soil', *Ambio*, vol 26, pp563–570

Cavalier-Smith, T. (1998) 'A revised six-kingdom system of life', *Biological Reviews*, vol 73, pp203–266

Cavalier-Smith, T. (2004) 'Only six kingdoms of life', *Proceedings of the Royal Society of London Series B*, vol 271, pp1251–1262

Daily, G. C. (ed) (1997) *Nature's Services*, Island Press, Washington, DC

Davic, R. D. (2003) 'Linking keystone species and functional groups: A new operational definition of the keystone species concept', *Conservation Ecology*, vol 7, issue 1, response 11

Hole, F. D. (1981) 'Effects of animals on soil', *Geoderma*, vol 5, pp75–112

Huising, E. J. (1993) 'Land use zones and land use patterns in the Atlantic Zone of Costa Rica; A pattern recognition approach to land use inventory at the sub-regional scale using remote sensing and GIS applying an object oriented and data-driven approach', PhD Thesis, Wageningen Agricultural University, The Netherlands

James, T. Y., Letcher, P. M., Longcore, J. E., Mozley-Standridge, S. E., Porter, D., Powell, M. J., Griffith, G. W. and Vilgalys, R. (2006) 'A molecular phylogeny of the flagellated fungi (Chytridiomycota) and a proposal for a new phylum (Blastocladiomycota)', *Mycologia*, vol 98, pp860–871

Jones, C. G., Lawton, J. H. and Shachak, M. (1994) 'Organisms as ecosystem engineers', *Oikos*, vol 69, pp373–386

Kibblewhite, M. G., Ritz, K. and Swift, M. J. (2008) 'Soil health in agricultural systems', *Philosophical Transactions of the Royal Society Series B*, vol 363, pp685–701

Lavelle, P. (1996) 'Diversity of soil fauna and ecosystem function', *Biology International*, vol 33, pp3–16

Lavelle, P. and Spain, A. V. (2001) *Soil Ecology*, Kluwer Academic Publishers, Dordrecht

Lavelle, P., Bignell, D. E., Austen, M. C., Brown, V. K., Behan-Pelletier, V., Garey, J. R., Giller, P. S., Hawkins, S. J., Brown, G. G., St John, M., Hunt, H. W. and Paul, E. A. (2004) 'Connecting soil and sediment biodiversity: The role of scale and implications for management', in D. H. Wall (ed) *Sustaining Biodiversity and Ecosystem Services in Soils and Sediments*, Island Press, Washington, DC

Lavelle, P., Bignell, D. E., Lepage, M., Volters, V., Roger, P., Ineson, P., Heal, W. and Dillion, S. (1997) 'Soil function in a changing world: The role of invertebrate ecosystem engineers', *European Journal of Soil Biology*, vol 33, pp159–193

Lavelle, P., Decaëns, T., Aubert, M., Barot, S., Blouin, M., Bureau, F., Margerie, P., Mora, P. and Rossi, J-P. (2006) 'Soil invertebrates and ecosystems services', *European Journal of Soil Biology*, vol 42, ppS3–S15

Lawton, J. H. (1996) 'The role of species in ecosystems: Aspects of ecological complexity and biological diversity', in T. Abe, S. A. Levin and M. Higashi (eds) *Biodiversity, an Ecological Perspective*, Springer, Berlin

Lawton, J. H., Bignell, D. E., Boulton, B., Bloemers, G. F., Eggleton, P., Hammond, P. M., Hodda, M., Holt, R. D., Larsen, T. B., Mawdsley, N. A., Stork, N. E., Srivastava, D. S. and Watt, A. D. (1998) 'Biodiversity inventories, indicator taxa and effects of habitat modification in tropical forest', *Nature (Lond)*, vol 391, pp72–76

Margulis, L. and Schwartz, K. V. (1998) *Five Kingdoms – An Illustrated Guide to the Phyla of Life on Earth*, 3rd edition, W. H. Freeman, New York

Moreira, F. M. S., Siqueira, J. O. and Brussaard, L. (2006) *Soil Biodiversity in Amazonian and other Brazilian Ecosystems*, CAB International, Wallingford

Rappé, M. S. and Giovannoni, S. J. (2003) 'The uncultured microbial majority', *Annual Review of Microbiology*, vol 57, pp369–394

Setäla, H., Laakso, J., Mikola, J. and Huhta, V. (1998) 'Functional diversity of decomposer organisms in relation to primary production', *Applied Soil Ecology*, vol 9, pp25–31

Stork, N. E. and Eggleton, P. (1992) 'Invertebrates as determinants and indicators of soil quality', *American Journal of Alternative Agriculture*, vol 7, pp23–32

Swift, M. J. (1998) 'Towards the second paradigm: Integrated biological management of soil', in J. O. Siqueira, F. M. S. Moreira, A. S. Lopes, L. G. R. Guilherme, V. Faquin, A. E. Furtini Neto and J. G. Carvalho (eds) *Soil Fertility, Soil Biology and Plant Nutrition Interrelationships*, SBCS/UFLA/DCS, Lavras, Brazil, pp11–24

Swift, M. J., Heal, O. W. and Anderson, J. M. (1979) *Decomposition in Terrestrial Ecosystems*, Blackwell, Oxford

Swift, M. J., Izac, A. M. N. and van Noordwijk, M. N. (2002) 'Biodiversity and ecosystem services in agricultural landscapes – are we asking the right questions?', *Agriculture, Ecosystems and Environment,* vol 104, pp113–134

Tittonell, P., Vanlauwe B., de Ridder, N. and Giller, K. E. (2007) 'Heterogeneity of crop productivity and resource use efficiency within smallholder Kenyan farms: Soil fertility gradients or management intensity gradients', *Agricultural Systems*, vol 94, pp376–390

Vanlauwe, B., Tittonell, P. and Mukalama, J. (2006) 'Within-farm soil fertility gradients affect response of maize to fertilizer application in western Kenya', *Nutrient Cycling and Agroecosystems*, vol 76, pp171–182

Wall, D. H. (ed) (2004) *Sustaining Biodiversity and Ecosystem Services in Soils and Sediments*, Island Press, Washington, DC

Woese, C. R., Kandler, O. and Wheelis, M. L. (1990) 'Towards a natural system of organisms: Proposal for the domains Archaea, Bacteria, and Eucarya', *Proceedings of the National Academy of Sciences USA*, vol 87, pp4576–4579

Chapter 2

Sampling Strategy and Design to Evaluate Below-ground Biodiversity

E. Jeroen Huising, Ric Coe, Juvenil E. Cares, Julio N. Louzada, Ronald Zanetti, Fátima M. S. Moreira, F.-X. Susilo, Suleymane Konaté, Meine van Noordwijk and Shiou P. Huang (deceased)

Introduction

Much of this handbook is concerned with how to take measurements once you are at a sampling site, a predetermined point in the landscape. Much of this chapter, however, deals with how the location of those measurement sites is determined. As will be discussed below, the problem of choosing the location of measurement points is one that occurs at several different scales. At one scale, we have to choose where the whole study will be located. At another, we need to choose where at an individual sampling site cores for soil chemical analysis will be taken. Somewhere between the two is the problem of choosing the sites themselves. While the argument gets a little more complex than this, the problem can be visualized as choosing the number and location of points in the study landscape at which the measurement protocols (described in subsequent chapters) will be implemented.

The problem is one of sampling. There is a long tradition of sampling in field ecology, and hence much experience has been accumulated. In addition there is a well established theory of sampling which applies anywhere it is attempted (Cochran, 1977). There are numerous texts describing both theory and application (e.g. Southwood and Henderson, 2000; Gregoire and Valentine, 2007). So why is another chapter discussing sampling for below-ground biodiversity needed? Despite the body of knowledge and experience we have, in any project there will be intense and sometimes divisive discussion of the sampling strategy. There are a number of reasons for this:

1 Application of the full theory, or of methods as extensive as those successfully used in other studies, may conflict with the practical constraints of the study being

designed. For example, it may not be feasible to take as many samples as you would like due to the time and cost, or access to ideal sampling locations may be restricted.

2 Application of sampling theory may require information that is not known until the data have been collected. For example, the sample size required depends on the variation between samples. If similar data have not been collected before, then this variation is not known.
3 There may be limits to the theory. More importantly, there are common misunderstandings of some of the basic principles, such as why random sampling works or what is meant by replication.
4 The objectives of the study drive the design. However these may not be fully developed, or there may be multiple objectives which suggested different approaches to sampling.
5 Scientists take differing philosophical stands on approaches to sampling, with a dichotomy between those who aim to 'See what is there, then seek to understand it' and those who 'Start with a hypothesis and seek to test it'.

In this chapter we describe some of the options for sampling for below-ground biodiversity, and the advantages and disadvantages of different approaches. This is followed by an example of the approach used in the CSM-BGBD project.

STUDY OBJECTIVES AND SAMPLING BASICS

Most authors writing on research study designs emphasize the point that the design is determined by the objectives. Kenkel et al (1989) explain this clearly in the context of ecological sampling. Many of the debates about appropriate sampling methods turn out to be due to differences of opinion as to the *exact* objectives of the study. Ford (2000) provides a comprehensive discussion of research objectives and approaches in ecology.

Simple random sampling (SRS) is the starting point for discussions on sampling. If the objective is to estimate population mean (such as the mean biomass of beetles per m^2 within the study area, or the mean number of fungal species within $1cm^3$) then SRS has relevant properties. The mean of the sample is an unbiased estimate of the population mean, and its standard error can be estimated without making any assumptions about the variation within the population (technically, a design-based estimate of sampling error is available). It is also intuitively appealing. Standard theory then shows how the precision of the estimate can be controlled by choice of sample size, and the precision increased (for a fixed sample size) by stratification. A useful alternative to SRS is systematic sampling on a grid, discussed in the section on random and systematic sampling.

However, few ecological surveys have the limited objective of estimating such a population mean. An example of an objective that requires a very different approach to sampling is that of inventory. If the aim is to identify all the species of a given group occurring in the study area then SRS is not appropriate. Think of a rare niche in the landscape (e.g. the bank of a pond that falls on the boundary between forest and field). There will be a tiny proportion of the whole study area occupied by such niches, so if we are trying to estimate the mean beetle biomass it does not matter if such locations are omitted from the sample. But they may well be home to species found nowhere else in the area, and hence should be included when the objective is inventory.

Most studies, including those of the CSM-BGBD project, aim to understand patterns in biodiversity. One approach to sampling is to collect data by SRS or grid sampling, seek to describe the patterns (for example by clustering and ordination) and then to explain them (for example, finding correlations with environmental variables). The alternative is to formulate some hypotheses predicting and explaining patterns in biodiversity, then design a study specifically aiming at testing the hypotheses. Proponents of the first approach may claim that they do not want to be 'biased' by initial hypotheses, or have their imagination and potential discoveries constrained by starting out with a limited objective. They would rather 'keep an open mind' and see what they can see. Of course important discoveries in ecology have been made by chance rather than through planned studies, and every scientist should permanently be open to the possibility of unanticipated observations, and truly novel explanations. But there are three reasons for trying to design a study with specific objectives, including testable hypotheses.

1. Those serendipitous discoveries usually have the nature of hypothesis formulation – observations which suggest explanations. Carefully planned studies are needed to test the explanations.
2. In fact the proponents of the 'no hypothesis' approach do have some hypotheses, but these are implicit. For example, without some notion of environmental factors that might be controlling biodiversity, it is impossible to choose which of an almost infinite number of such factors should be measured at sample locations. If the implicit hypotheses are made explicit, study designs can be improved.
3. If we have specific hypotheses, it is often possible to improve the study design, making the study more efficient.

The last point underlies much of what follows in this chapter. Suppose the hypothesis is that below-ground biodiversity in agricultural plots is determined by the level of disturbance (D) and the level of soil organic matter (SOM). If we collect data by SRS or grid sampling, then it is likely that:

- Most sample locations will have values of D and SOM around the average, with relatively few points with very high or low values. But when seeking to understand the relationship between below-ground biodversity and SOM or D, it is the more extreme points that provide most of the information (Figure 2.1). Stratification can be used to increase the number of points with more extreme SOM and improve the estimate of the relationship without increasing the number of samples.
- SOM and D may well be correlated, for example with plots with high D typically having low SOM. Then it is hard or impossible to disentangle the effects of the two variables. However, the study could be designed to deliberately include some samples with high D and high SOM, and some with low D and low SOM. Then the effects of both variables, and their combined effect, can be estimated.

In practice, it may not be possible or useful to produce a single index of below-ground biodiversity, as plotted in Figure 2.1, and relationships may be more complex than straight lines, but the same principles of design apply.

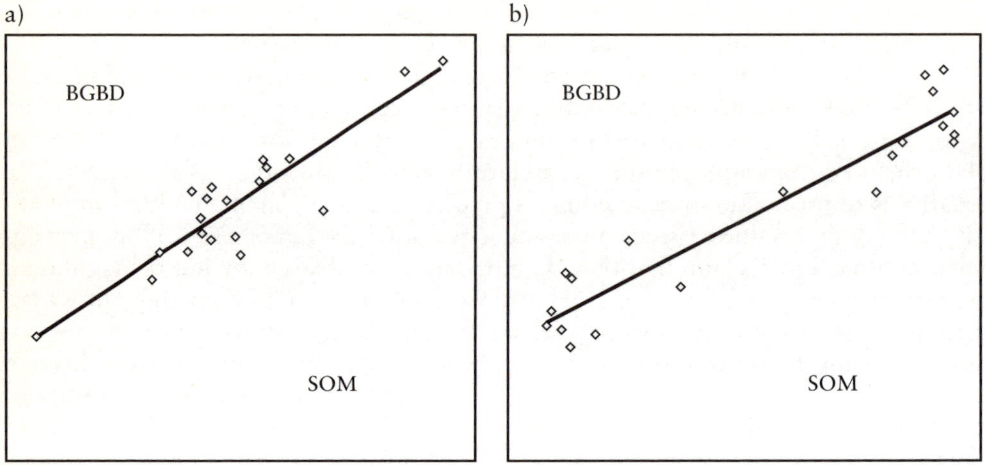

Figure 2.1 *Designs for estimating the relationship between below-ground biodiversity and SOM: a) simple random or grid sampling will probably give most SOM values near the average, and a poor estimate of the line; b) deliberately including samples with more extreme SOM values through stratification increases the precision of the estimate of the line at no extra cost*

Another example of a hypothesis implicit in many studies is that of the spatial scale at which interesting patterns occur. By choosing the distance between sample locations and the overall size of the study area, the scientist is making choices and assumptions about the important scales to study. If these are made explicitly then they are open to others to debate, with a likely improvement in the study design.

The overall objective of the CSM-BGBD project was to test the hypothesis that increasing land use intensification changes below-ground biodiversity (more specifically leads to a loss of below-ground biodiversity). This is a rather general statement, but can still be helpful in focusing design of sampling. Ideally we would investigate it with an experiment. The only certain way to determine the effect of changing something is to change it, and that is the basis of an experimental approach. However, this is often not feasible. If we have to use an observational study design, rather than experimental, then the ideal would be a longitudinal study, in which plots are monitored over time to see whether changes in below-ground biodiversity can be correlated with changes in land use. This generally is also not feasible within a project of fixed and short duration, as the time over which monitoring may be needed is not known. Hence the study, like most others, used a cross-sectional approach, looking at a range of land uses at one time point. The hope is that correlations between land use intensification and current below-ground biodiversity do reflect some causal connections, and give indications of what would happen to below-ground biodiversity if land use changes take place in the future. Though the validity of this approach can be questioned, it is often the only option available. Discussions in this chapter therefore only consider alternative sampling schemes for collecting cross-sectional data.

PRACTICAL APPROACHES

Designing a successful, practical sampling scheme is an art. It requires deep understanding of the scientific basis of the research and of the properties of alternative methods. But these need blending with the practical constraints imposed by cost, time and expertise available. There may well be additional constraints, for example limited access to desirable sample locations, or the need to rapidly transport samples from field to laboratory. The details of how these practical and theoretical sides can be merged will be different for every study, and give each investigation its own unique aspects. However, it is possible to set down outline steps in the process that can be followed in any study.

Step 1 Define objectives

As outlined in the section on study objectives and sampling basics, the objectives determine all aspects of the design. Hence they must be clearly and precisely determined at the start. Write them down, so that it is easy to share them with others for suggestions. Get comments and suggestions from as many other scientists as you can. These could be scientists working on similar topics but in other locations, those who have worked in the same location or those with experience in the methods you will use.

One tool to help refine objectives is the simulated presentation of results. Imagine you have completed the study and obtained results. What tables and graphs would you like to be able to present, in order to meet your objectives and provide evidence for your hypotheses? Write these down, with realistic numbers and patterns – Figure 2.1 is a simple example. Then check carefully that a) those results really would meet the objectives, and b) that the sample design imagined really could give those results.

Step 2 Review other studies

Look at reports from other related studies. While each study has some unique aspects, there are always earlier studies from which you can learn. Try to understand which aspects of the methods used appeared successful, and which ones seemed to limit the efficiency or quality of results. Note in particular sample sizes used and the variability in results.

Step 3 Assemble background data

Assemble background information that will be needed to design sampling details. These include topological maps (for example to stratify by altitude or understand access problems), remote sensing images (for example to map ground cover), land use maps (for example to identify the main land uses to include in the study), meteorological data (for example to help decide on suitable seasons for field sampling).

Step 4 Produce a design

Produce a tentative design using a combination of general principles, your own experience, designs used in other studies and imagination. There may well be aspects you have very little idea about, but make a realistic suggestion. Write the design down with as much detail as possible.

Step 5 Review the design

Give the design to other scientists to review and make comments. Again, these may be people who have worked on similar topics, used similar methods, worked in the location or are just generally perceptive. Include a statistician with experience in ecological research. A statistician is likely to see aspects of the problem that might be missed by ecologists.

Step 6 Pilot

Try out the approach. A pilot investigation is a chance to evaluate the practicality of the sampling scheme. It also allows testing and refinement of measurement protocols, data handling procedures and so on. It allows estimation of the time taken to sample and process samples. If it is possible to process some measurements to the point of statistical analysis, then it also gives an indication of variability, which can be used to help decide final sample sizes.

Step 7 Iterate

At any step, expect to go back to an earlier step and try again. In particular, revise objectives in the light of new information and insights. A common mistake is to get information that suggests the objectives are unobtainable, but to carry on anyway.

HIERARCHY, REPLICATION AND SAMPLE SIZE

Most study designs are hierarchical and the sampling problem is not simply one of selecting measurement locations within a study area. The CSM-BGBD project provides a good example. It involves several countries. Within each country one or more benchmark locations were selected. In each benchmark, one or more study areas (labelled 'windows') were selected. Within each study area about 100 sample locations were selected. At each sample location measurements are taken. The measurement protocol defines further layers in the hierarchy, such as four cores being taken for soil characterization, and subsamples of the cores subject to chemical analysis.

At each layer in the hierarchy the basic sampling questions recur: how many units should be selected, and which ones? At the highest levels the answers may not be based on scientific grounds. Selection of countries may be based on politics or the interests of funding agencies and researchers leading the project. But at some level selection should be based on the objectives of the study and application of some principles.

The first is the sampling theory idea of a 'population' to be sampled. The terminology is confusing, as this has nothing to do with a biological population. The notion is one of knowing what your results will refer to. As an example, we could study belowground biodiversity on farms around the forest boundary of Mt Kenya. That would require a sample of farms from that location. If we wanted results that apply to the forest boundaries on mountains in East Africa generally, then we need to sample from some of the other mountains as well. Without that, we can only make statements about Mt Kenya on the basis of the data, with extrapolation to other locations dependent on other information or assumptions. The implication for sampling is that the overall area about which we want to makes inferences (the 'population') needs delineating before a sampling scheme can be determined.

The second idea is that of replication, which concerns consistency of patterns and relationships. The aim of research is to find some patterns, such as patterns of below-ground diversity related to land use. Patterns of interest are those which are consistent across a number of cases, as it is only these that can be used for prediction and may reflect some underlying rules or processes. Hence we need repeated observations to determine whether patterns are indeed consistent. Suppose we have ten samples taken from a forest and ten from nearby cultivated fields, and the forest plots consistently have higher below-ground biodiversity. What can we conclude? If the samples were selected appropriately, we can conclude (to a known degree of uncertainty, assessed by the statistical analysis) that the forest plots are more diverse than the fields. But strictly speaking, we can only conclude that *that particular forest* is more diverse, not forests in general. If we seek a more general conclusion, then we should look for consistency across several forests.

Within a hierarchical study design, higher-level units such as benchmark sites may provide one level of replication and consistent patterns across benchmark sites probably represent some widely applicable 'rule'. But within benchmark sites we would make stronger conclusions if we ensured that several, rather than a single, forest (or other land use element) is sampled. Multiple samples from the same forest may not serve the same purpose, representing 'pseudo-replicates'. The extent to which repeated samples within one forest serve the same purpose, or can be interpreted the same way as samples from different forests, depends on properties of the data and not of the design. The safe approach is to ensure valid replication and some generic results by a design that replicates forests and other land use elements.

Some of the implications of these ideas for grid sampling are illustrated for a simplified example in Figure 2.2. The aim is to sample a landscape with two land uses, labelled 'forest' and 'agriculture', in order to examine differences in below-ground biodiversity. In Figure 2.2a, a single large grid has been laid down in such a way as to include both land uses. The grid is a single 'window' with 77 sampling locations (intersections) defined. In Figure 2.2b, three smaller windows are used in order to sample three

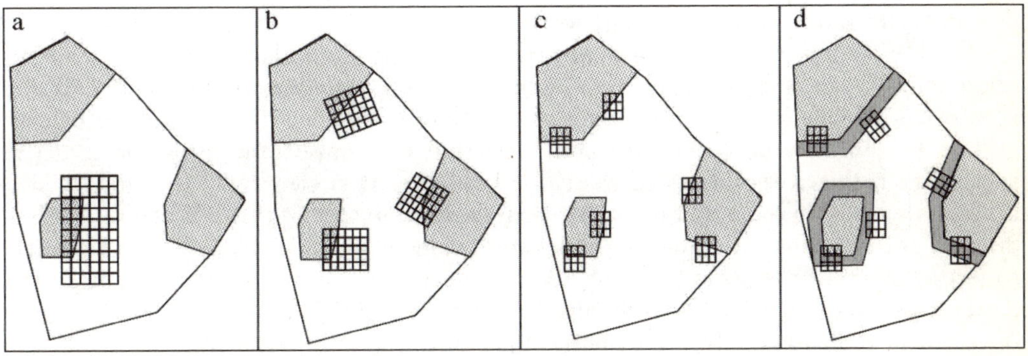

Figure 2.2 *Four approaches to using grid sampling in a landscape with two land uses, forest and agriculture: a) a single grid that includes one forest patch; b) three grids that sample three different forest patches; c) increasing the replication; d) recognizing the boundaries as another category*

different forest patches, rather than one only. The replication can be further increased, and more of the whole study area observed, by using more, smaller windows (Figure 2.2c). One criticism of this third design is that all the sample locations in agricultural land fall close to a forest boundary, and may not be considered to be representative of the land use. A response to this is to define a new category of 'forest boundary' and ensure that windows sample all three (Figure 2.2d). Notice that it is not necessary to have all land uses sampled in each window. If this process of reducing the size of windows while increasing their number is continued, then eventually we lose the possible advantages of grid sampling, and end up with a design that looks like a random sample of individual locations.

'Scale' is a confusing and controversial idea in ecology (Peterson and Parker, 1998), but it is clear that the scale at which we anticipate (or hypothesize) patterns determines the level in the hierarchy at which replication is required. For example, the hypothesis may be 'below-ground biodiversity in agricultural plots decreases with increasing distance from the forest edge'. This can be investigated with plots (sample locations) at a range of distances, with replication of each distance. A different hypothesis is 'below-ground biodiversity in agricultural plots decreases with decreasing forest cover in the landscape'. Here we need to define what is meant by 'in the landscape' – that is, the spatial scale at which forest cover is assessed. Suppose that was defined as areas of $1km^2$. Then the hypothesis needs a sample of $1km^2$ units, with varying levels of forest cover. Replication now dictates the need for several such units at each level of forest cover. To assess the below-ground biodiversity within such a $1km^2$ unit will require further sampling, with definition of some sample locations within each unit. The replication at the within-unit level is important for determining the precision with which the below-ground biodiversity for each unit is measured, but it is not relevant to affirming the consistency of pattern across $1km^2$ units needed to examine the hypothesis.

In other areas of ecology, landscape factors (e.g. forest fragmentation) are found to affect processes, so objectives of a below-ground biodiversity project may include 'landscape analyses'. The two examples in the previous paragraph are both examples of analyses that use landscape factors, yet are based on data from different levels in the hierarchy, one using plot-level data and the other data from $1km^2$ units. The message is clear: 'landscape level' is not well defined, and aiming to do a 'landscape level analysis' does not tell you the sampling design needed.

Once we know what is to be replicated, standard methods are available to help select sample size, and software to implement them is widely available. The methods require knowledge of two things: the magnitude of difference (for example in below-ground biodiversity between two land uses) that it is important to detect, and the variability between replicates of the same land use. It is clear why the sample size decision depends on these, but it is usually rather hard to specify them. When research is directed at measuring economic responses to management decisions (e.g. crop response to fertilizer) then it is feasible to specify a minimum response that it is important to detect. However when the research aims to detect and understand some processes, it is often not possible to specify a size that is important.

A rough estimate of variance can often be obtained from previous studies, but how relevant these will be in new environments may be unknown. Another complication arises from the multivariate responses of interest. The standard methods assume there is a measured response of below-ground biodiversity that we can use when planning sample size. But any real study has multiple responses of interest, such as the diversity of different functional groups measured in different ways, numbers, biomass and ratios of these for functional groups or even species, and so on. Hence in practice sample size has to be based on a combination of information from formal methods (which can give indications of orders of magnitude needed), previous similar studies, and pilots.

Sequential approaches (Pedigo and Buntin, 1993) allow sampling to continue until some criteria are met. While theoretically attractive, they are unlikely to be practical for below-ground biodiversity studies as work needs planning in distinct phases of field and laboratory, with many measurements only becoming available a long time after field sampling. Adaptive sampling (Thompson and Seber, 1996) allows the design to respond to patterns being detected. Again there are some attractions in the idea but they are unlikely to be feasible given the need to plan field campaigns in advance. A range of multi-scale designs have been used in ecological studies. The idea of these designs is to choose sampling positions so that patterns at several different scales can be investigated. Fine-scale patterns require points close together. Larger-scale patterns require points further apart. Hence both are included, with efficient designs having a clustered structure (Urban et al, 2002; Stein and Ettema, 2003).

At each selected sampling location, further sampling is needed (see 'Point measurement scheme' below). Think of the selected location not as a point but as plot, perhaps with an area of the order of $100m^2$. Sampling is needed within this plot, as only a very limited volume of soil can actually be examined for most below-ground biodiversity measurements, and several samples are taken in order to represent the whole plot. However, typically the measurements within each plot are bulked – that is, the several soil samples from the plot are mixed before measurement of the below-ground biodiversity. There are two reasons for bulking. One is simply practical. There would be too many samples to process without bulking. The second is the need for coincident measurements of different functional groups of below-ground biodiversity. If several groups of species are being assessed, then presumably the relationships between them are important. This means they must be measured in the same place. However it is usually only possible to examine one group in a given soil sample, and extracting the sample for one group may disturb it for others. Hence all measurements are at the plot level. This means that variation and patterns at the scale of within-plot (e.g. <10m) are not examined.

FOCUS ON OBJECTIVES: STRATIFICATION

In the introduction to this chapter we suggested that a focus on objectives of a study will increase the efficiency of the design, and increase the chance of meeting the objectives. Objectives of discovering and understanding land use effects on below-ground biodiversity require comparison of different land uses. One approach to improving the sampling design is to use 'stratification' to ensure that we do indeed have adequate sample sizes of each land use. Used in this sense, the strata are land areas under different uses, and the idea is to deliberately sample from each of these. It is sometimes suggested that this approach is 'biased', as the land use classes to sample are determined *a priori*. If the data were used to make statements about the overall study area (e.g. the mean number of beetles per m^2) without accounting for the design, then the result may be biased, as different land uses may not be represented in the sample with frequencies that are proportional to their occurrence in the study area. But the design is not biased for the objective of comparing land uses. Furthermore it is efficient. If we have a total of N samples to compare two land uses then, in the absence of any further information, the best design is to have N/2 in each of the two groups. With the stratified sampling approach we can choose a suitable sample size for each land use.

If this approach is to be employed then there are two prerequisites:

1 We need to know which land uses will be compared and have precise definitions of them.
2 The location of these land uses must be known – a land use map of the study area is needed.

The first of these makes some scientists uncomfortable, with the feeling that prior definition of the land uses to investigate excludes discovery of potentially important patterns. But the definition has to be done at some stage anyway. The need to define them precisely also has to be done at some time. For example, where is the boundary between 'pasture with trees' and 'secondary forest', when we move along a gradient of increasing tree cover? Here we have another potential gain in efficiency from thinking through these requirements at design rather than only at the analysis stage. If a sampling design does not take land use into account, then there is a good chance that many of the selected sample locations will end up in positions of ambiguous land use definition that we are not sure how to classify. With the stratified approach, these areas can be excluded from the sampling. Of course, if the aim is inventory of the landscape, we do not want to exclude some land use types and those transition zones may be important. But if the aim is investigating land use effects it does make sense to exclude such locations.

If the objectives include investigation of boundaries between areas of different land use, or of rare niches such as linear features, then these should be specifically included in the sampling. Without doing so it is likely that the sample will include only a few observations of these categories from which nothing can be concluded. It is much more efficient to either include them with a large enough sample size if they are required by the objectives or exclude them (give them a sample size zero) if they are not required by the objectives. Note that much the same arguments apply if the hypothesized factors influencing below-ground biodiversity are not land uses per se, but environmental variables influenced by land use, such as SOM or frequency of fire. The requirement to have land use mapped to be able to use it in sampling should not be a constraint, with interpretation of remote sensed (RS) imagery a possibility if other land use maps at suitable resolution are not available. The same may not be true if variables such as SOM are to be used for stratification. It may be useful to do a rapid survey of SOM, calibrate it to a land use map or RS image, and use that to define strata.

Random and systematic sampling

The essential reasons for using SRS in many applications were outlined above and are elaborated in texts such as Cochran (1977). To implement SRS it is necessary to delineate the study area then select sampling locations inside it at random, in such a way that a) every point is equally likely to be selected, and b) selection of one point does not change the probability of including any other point. Stratified random sampling requires doing the same thing within each stratum. With software to aid in the randomization, and GPS to locate selected sample locations in the field, this scheme is feasible in practice. However, ecological sampling often uses non-random sample selection, sometimes with good reason.

A common non-random approach is subjective selection of sample locations. This means, for example, choosing the samples to include sites judged to be interesting or important, and is often the basis for selecting sampling units at higher levels in the hierarchy. While sometimes necessary, the limitation of this approach is that the 'representivity' of the sampled area (the extent to which findings can reasonably be assumed to apply to a larger population) depends on the judgement of the designer, not on any inherent property of the design. It is therefore open to dispute when results are presented. If a subjective sample of size 1 is taken, this is equivalent to limiting the study area. For example, if a single 'window' is subjectively placed in a benchmark area, then in fact we have reduced the study to that window, and any claim to represent the benchmark area depends solely on the expertise of the designer.

Systematic sampling has found much application in ecology, both with one-dimensional transects and two-dimensional grids. In the case of transects, samples are selected at points a fixed distance apart along a predetermined line. For grid sampling, a (usually) rectangular grid is defined in the area, and samples taken at each intersection point. The potential advantages of these types of systematic sampling derive from both theory and practice. The practical advantages include:

- Ease of locating sampling points, and describing the location and means of finding them to field teams. For example, the protocol may be something as simple as 'From the starting point, walk north and sample every 50m'.
- Ease of planning fieldwork, for example estimating the time needed to sample a fixed number of points.

The statistical reason for using them is that they can be efficient (Webster and Oliver, 1990). Consider a study with the objective of measuring the average or total of some quantity (for example total soil carbon in the study area or average number of beetles per m^2). A grid sample will give a better estimate than a simple random sample of the same size if the measured quantity varies in a patchy way, which is typical for environmental and biological variables. The efficiency comes from the fact that closely neighbouring points are similar to each other, so do not add much new information, and the grid spreads the sample as evenly as possible through the study area. For similar reasons, the grid approach can be expected to be good for compiling the inventory of a study area except that it may miss rare niches (see below).

There are some negative aspects of grid sampling:

- Some points of the grid may be at points which should not be included in the study, such as roads or water bodies. Obviously these must be excluded.
- Grids will sample different land uses with a sample size roughly proportional to the areas of those different land uses. In particular, rare land use classes may be omitted completely. While this can be compensated by moving the window around and adding points, the process could be rather arbitrary and subjective.

- It is sometimes not possible to characterize the land use unambiguously at every sample point.

These are all related to the problem discussed in 'Focus on objectives: Stratification'. If the aim of the study is comparison of land use classes, then grid sampling may not capture those in an optimal way. Thus grids and transects are probably most appropriate for sampling below-ground biodiversity when either a) there is no explicit objective or hypothesis involving comparison or relationship with environmental variables, or (b) the hypothesis refers to a higher-level spatial unit than the scale at which the grid or transect sampling is done. For example, Swift and Bignell (2001) recommend 40m-long transects, but these are *within* each land use class. For the purpose of comparing land uses, transects are replicated and randomized to strata defined by different land uses.

In this way, systematic grid or transect sampling are usually combined with random sampling. For example, there may be several grids defined, as in Figure 2.2d, with their location and orientation randomized. Similarly the starting points and orientation of repeated transects may be randomly selected. Transects can also usefully be aligned with environmental gradients hypothesized to be important, when they are known as 'gradsects' (Wessels et al, 1998). With randomization at some level in the hierarchy, statistical analysis based on the random properties of the design is possible. For example, if a number of small grids are randomly placed in the study area, then we have the replication necessary to establish the consistency across windows of patterns found.

Statistical analysis at the sample point level cannot be based on randomization, as the locations were not independently selected within each grid. There are two possible approaches to analysis. One is to assume that the data behave at random (i.e. the statistical properties are the same as if the point had been randomly located). The second is to use an explicit model of spatial pattern. In most analyses looking for relationships between environmental variables and below-ground biodiversity the former method is used, mainly because alternatives are complex. The consequences of this assumption are rarely investigated.

It is clear that the spacing and overall size of a grid determine the scale of patterns that can be detected with it. It will not be possible to pick up patterns (e.g. patchiness in below-ground biodiversity) at spatial scales less than the distance between points in the grid. Likewise, it will not be possible to detect patterns larger than the overall size of the grid. In fact, the maximum pattern size must be less that the size of the grid, as the patterns can only be recognized if there are several repeats within the grid. It is this aspect of pattern scale, set by the objectives of the study, that should determine the spacing and overall size of a grid.

It is sometimes suggested that grid spacing should be such that neighbouring points are uncorrelated. This notion of spatial correlation is important but also confusing. The correlation between measurements a given distance apart is not an absolute quantity, but is measured relative to an average (technically, the issue is one of stationarity). To see this, think of analysing data from a single window in Kenya. Points more than 200m apart may well show no similarity in below-ground biodiversity. But if we put data from a global dataset together we would expect to find similarity not just between points in the same window but perhaps between all points in Kenya.

DEALING WITH VARIABILITY

Experience from past studies suggests you should expect a high level of variation in many key measurements of below-ground biology. Even over short distances we expect large variation in numbers and diversity of different functional groups. In tropical agricultural landscapes the variation within a land use category may be considerable in terms of management practices, variation in above-ground vegetation characteristics, difference in land use history of the plot, edge effects, topographic position and bio-physical characteristics. If formal methods of determining sample size requirements were followed through, they would be likely to give indications of sample size many times larger than that which is feasible and affordable. What should be done?

First, there is no point in doing nothing! Simply carrying on with the preconceived sample size will mean objectives will not be met. If the original plan was to have about ten samples of each land use within a benchmark site, and the indications are that we need about 100 samples of each, there is no point continuing. You will end up with vague and inconclusive results, reflected in high standard errors and no significant effects when analysing the data. There are three possible responses:

1 Increase the sample size.
2 Use sampling methods to reduce the variability.
3 Reduce the scope of the study.

The first option is obviously impractical in many cases. There are always limitations in time, money, facilities and expertise.

There are various methods of reducing variability by sampling. Most useful are ideas of stratification and matching. Note this use of the term 'stratification' is the one common in sampling, but different from that in 'Focus on objectives: Stratification' (above). If some sources of variability can be predicted, then they can be used to define strata and removed from the analysis. For example, if the benchmark site covers a range of altitudes, we may expect variation in below-ground biodiversity by altitude. Stratification would then divide the site into altitude zones, and sample within each of these. At the stage of data analysis, land uses would be compared within strata and the between-stratum variation would not obscure the results. This approach requires that some (not all) different land uses occur within altitude zones. If land use only varies with altitude then the two factors are confounded and their effects on below-ground biodiversity cannot be distinguished. It is typical for environmental variation to be patchy, and hence for some of the variation in response to show patchiness. Hence strata may be usefully defined as geographically close sets of sampling points. The windows in Figure 2.2 can be seen in this way.

Matching takes the idea of stratification to an extreme. Suppose two of the land uses to be compared are forest and maize fields. We can expect the below-ground biodiversity to depend on many environmental variables such as climate, topography, soil and geology. These environmental variables typically vary in a patchy way, with close sites being similar. Hence if we choose forest and maize plots that are close together, then differences between them will be mainly due to the land use rather than other factors, and we remove those other 'noise' factors from the analysis. Thus the approach would be to identify and sample, say, ten pairs of sites, each pair consisting of forest and maize plots that are close together, either side of a land use boundary. Formally, each pair constitutes a stratum of size 2. For more than two land uses the design can be extended. Ideas of design for incomplete block experiments are relevant to choosing suitable pairs of land uses to match.

Managing variability by reducing the scope of the study is often the best solution. The scope could be reduced by reducing the size of a benchmark site, naturally reducing the heterogeneity. This is unsatisfactory as it also reduces the generality of the result. If we only sample in a small area, then there is no basis for assuming we have found patterns that are widely applicable. Other ways of reducing the scope of the study are:

- not including all land uses found in the benchmark area, but a selection that covers a clear gradient in land use intensity, or represents some typical land use transitions;
- tightening the definition of a land use class. For example, rather than having 'maize field' as a land use, we could limit attention to maize fields that have been in continuous cultivation for ten years, have not received fertilizer in the last three years and are tilled by hoe;
- avoiding samples in ambiguous sample locations, such as those near a boundary.

While these will all help in detecting and measuring the effect of land use intensity on below-ground biodiversity, they may not be consistent with objectives of species inventory. A trade-off between these two objectives may be necessary. This is common in design, the bottom line being that we cannot expect to find out everything from one limited size sample.

POINT MEASUREMENT SCHEME

Once sampling points are located, a measurement protocol must be defined. That actually involves further sampling. The basic point sample scheme adopted by the CSM-BGBD is presented in Figure 2.3, complemented by Table 2.1. Though the point measurement scheme provides for sampling of the same functional groups that are described by the procedures of the Alternatives to Slash-and-Burn project (Swift and Bignell, 2001), the scheme is rather different from what is proposed by Swift and Bignell, to make it suitable for a grid-based approach. For example, the fewer monoliths per sampling point are offset by the larger number of sampling points, resulting in approximately the same number of sampling events. Nonetheless, preliminary results have indicated that with one small monolith per sampling point not all species may be captured. If that is the objective for the sampling, options are provided for additional larger monoliths to be included.

Macrofauna: a single monolith, a set of pitfall traps and at least one transect of 20m is deployed in each location sampled. The transect sampling is amplified to include ants and soil beetles, in addition to termites. In an alternative scheme additional monoliths can be dug to improve the sampling of earthworms, and a simple regime of casual sampling for termites is also introduced. Note that in small field plots, highly dissected cropping systems or on difficult terrain, it is not necessary for the transect to be linear. A

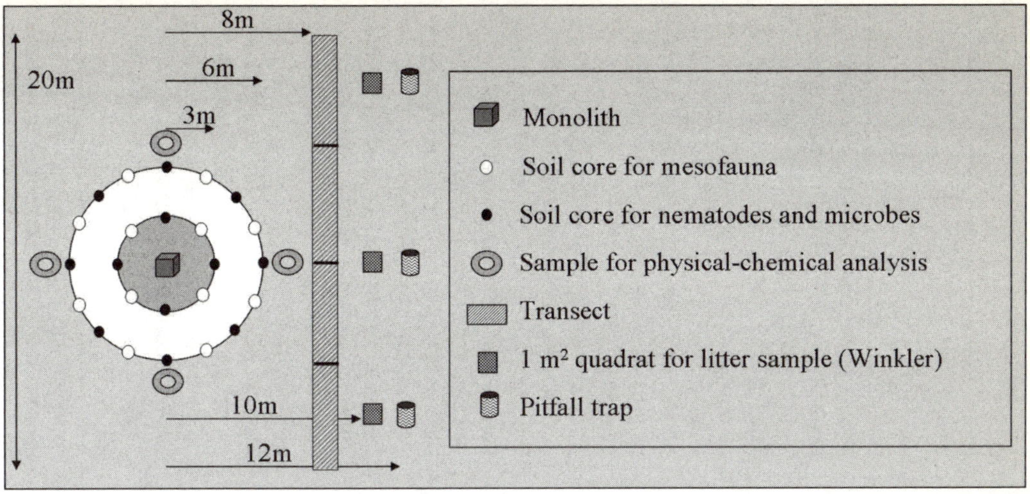

Figure 2.3 *Minimum point sampling scheme for all biota. Sampling can be extended by using one or two additional transects for termites, ants and beetles, by casual sampling for termites (one hour) and by additional monoliths to capture more earthworms*

transect can be bent through angles up to 90 degrees to sample plots of irregular shape or to avoid significant natural features such as streams, steep slopes or rock outcrops. Tree falls should, however, be included in the transect if this is appropriate to its existing line and length, and not bypassed (Swift and Bignell, 2001).

Mesofauna: 12 soil cores taken from points on two concentric circles around a single monolith, at 3 and 6m (radius), and extracted by the Berlese method. In addition, unbaited pitfall traps serve to collect active and surface-dwelling soil meso- (and macro-) fauna over a 24-hour period. Pitfalls are generally good for ants, some beetles, some grasshoppers, millipedes and spiders. Soil cores extracted by the Berlese method produce collembolans, acari (mites) and (in wetter sites) enchytraeid worms.

Microfauna: sampled similarly by 12 soil cores arranged in two concentric circles around a single monolith, at 3 and 6m (radius), from the monolith. 'Microfauna' refers principally to nematodes, although very small mites (and mite larvae) extracted from soil cores by the Berlese method can qualify.

Microsymbionts and soil fungi: sampled similarly by 12 soil cores arranged in two concentric circles around a single monolith, at 3 and 6m (radius), from the monolith.

Soil physical and chemical parameters: four soil cores located on a circle of 6m radius from a single monolith. For co-location studies, an additional soil core can be taken from the outside wall of the monolith trench (see Chapter 3 on macrofauna sampling).

In an alternative scheme (Figure 2.4), land uses can be sampled by a combination of principal and subsidiary monoliths, 50m transects, spaced litter quadrats and pitfall traps. Casual sampling in an envelope around the transects can also be employed and

Table 2.1 *Groups of soil organisms that are collected using the various contraptions*

Symbol of method	*Diversity component/groups of soil organisms* Ants	Termites	Beetles	Earthworm	Other macrofauna	Nematodes	Microbes	Mesofauna.	*Soil*	*Description*
Pitfall	✗	✗	✗		✗			✗		Pitfall trap or baited pitfall trap* – 3 traps per sampling point. Traps must be emptied 48 hours after being put in the field.
Transect		✗		✗						Semi-quantitative transect – 2×20m – 1 per sampling point, perpendicular to the slope.
Winkler	✗	✗	✗		✗			✗		Winkler extractor for litter sample (1×1m) – along side of the transects, 3 Winklers per sampling point.
Berlese	✗	✗	✗					✗		Soil cores with litter (3.5×3.5×5cm) – 12 subsamples per sampling point (depth 0–5cm). For mesofauna make 1 composite sample (1st option) or 4 composite samples composed of 3 subsamples of each quadrant of the circle (2nd option) or 3 composed of 4 subsamples.
(Various)						✗	✗		✗	Soil cores for microbes at 0–20cm depth without litter. Circles radius: 3 and 6m, 1 composite sample of 12 subsamples per sampling point.
Monolith	✗	✗	✗	✗	✗			✗		Monolith (25×25×30cm) – including litter – 1 per sampling point – TSBF method.
									✗	Soil sample – 1 sample composed of at least 4 subsamples for chemical and physical analysis (cores) at depths 0–10; 10–20; 20–30cm.

Note: * The suggested bait for the pitfall traps is monkey or human dung/dregs. Dung from humans or monkeys is very effective to attract dung beetles (Coleoptera) and at the same time it does not deter or interfere with capture of saprophytic beetles (Scarabaeidae, Aphodiidae, Trogidae) and predators (Histeridae, Staphylinidae). Omnivore dung is more suitable for tropical forests areas as omnivores dominate in those environments. Cow dung (herbivores) is more suitable for savannas or grassland areas. Baits might attract non-targeted groups like flies and ants.

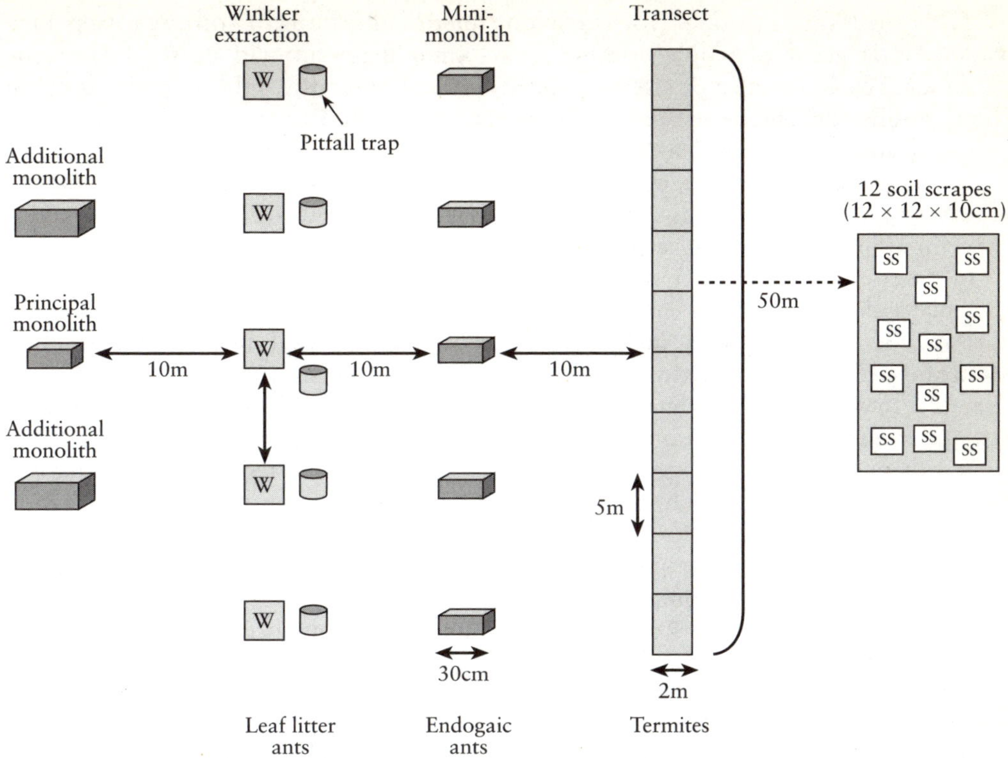

Note: Details of transect sampling procedures are given by Swift and Bignell (2001) and in Chapter 3.

Figure 2.4 *Alternative scheme for sampling macrofauna, making use of a 50m transect*

Malaise traps erected to catch the larger flying insects. The protocol differs from the one proposed above, as the transect points, litter quadrats, pitfall traps and subsidiary monoliths are placed in parallel 50m lines, with about 10m lateral separation. This protocol should basically be regarded as an extension to the one proposed above, and could be considered to apply where more importance is given to sampling all species that are occurring within the area and in cases when sampling frequency (number of sampling plots) is reduced for reasons explained earlier. These modifications are also more suitable where strip patterns of land use are encountered in all localities.

PRACTICAL CONSIDERATION IN ORGANIZING THE FIELD CAMPAIGN

Collection of samples or direct measurements in the field should as much as possible be completed at each location sequentially and during a single excursion. For soil physical and chemical parameters that remain stable over time (only), sampling can be permitted at a different point in time. This, of course, does not apply for highly dynamic features like soil water content.

The collection of samples for the main functional groups of soil organisms may involve a large group of people working at each sampling point, with a risk of damaging the standing crop, but also of affecting the inventory, especially of those organisms with a high motility or which are sensitive to human intrusion. For example, epigeic earthworm and some types of soil-feeding termites are sensitive to disturbance, and especially to the trampling of plots, and may move away very rapidly. It is therefore important to reduce the number of persons directly collecting samples, but still necessary to have a sufficient number on hand for efficient handling and processing.

In the training of technicians and field assistants, attention should be given to the particular risks that arise for each specific group of soil organisms. The risk of contamination when dealing with microorganisms should be well understood and the importance of proper procedures to avoid contamination should be impressed upon persons if they are less experienced. The same is true, indeed, for the risk of organisms escaping from sampling locations as a consequence of acute disturbance.

Another concern is the large volume of soil that will be collected during the sampling and the need to reduce the volume for transport. From this point of view the extraction of specimens from the samples is best done at or near the measurement sites. Also the bulking of samples and subsequent extractions from the bulked sample of subsamples for analysis should also be completed on site.

Storage temperature and storage time for samples is an important consideration, as many soil organisms may die at the ambient temperatures of the air or the surface of the ground, or if kept in closed bags or containers for more than a few hours. To deal with these risks it is important to identify a cool local storage facility in advance of the sampling, or else arrange for immediate on-site extraction of specimens. In some cases, for example the Berlese funnel apparatus for mesofauna and the Winkler bag system for extracting ants and beetles from litter, the equipment can usually be set up in the field laboratory for immediate use. In other cases, where organisms need to be cultured in vitro or where ambient soil properties need to be determined, it is advisable to establish a rapid transport link to the laboratory. Under no circumstances should samples be left out in the sun, even where the extraction protocol calls for eventual drying out of the material. Further details of these precautions and general logistical advice are provided in the chapters that deal with specific groups and methods.

Sampling in the CSM-BGBD project

The sampling hierarchy

In the project, benchmark areas were selected in seven countries in the tropics, with the aim of assessing the loss of below-ground biodiversity with increasing land use intensity. The benchmark sites were located in major biogeographic regions in the tropics, ranging from the Amazonian humid tropical rainforest to the Himalayan mountain forest of India, and representing areas of global biodiversity importance. Each benchmark area included a range of land uses from pristine, or relatively undisturbed old-growth forest to more intensively used agricultural lands, to determine whether common trends can be observed in the change of below-ground biodiversity in relation to intensity of use. The use of standard sampling methods across countries and benchmark areas allows comparison between sites.

An approach is adopted for sampling within benchmark sites that makes use of sampling windows. A window here is a (usually rectangular) parcel of land that includes the range of locally significant land uses for study and is considered 'representative' of the benchmark site. Windows are typically 2–5km^2. The sampling windows vary in size and number due to differing land use configurations in each benchmark area, and in one case because of access problems. For example, one larger window is used when that captures the relevant land uses in the area, but up to six smaller windows are distributed within the area to sample the targeted land uses as required.

A regular grid (with variable spacing, see below) is used to identify a set of possible sampling points within each window, the points corresponding to the intersections of grid lines. Grid sampling was used for the reasons, described above, of statistical efficiency and to rationalize fieldwork. Furthermore, the inventory is of a more exploratory nature, irrespective of the general hypothesis formulated, because questions remain as to which of the factors that determine land use intensity exercises the strongest influence on the distribution and richness of soil biota, and as to the scale at which the spatial distribution patterns manifest themselves. The approach is similar to the one used in the European Union funded BioAssess project, which had a similar objective of evaluating biodiversity in relation to intensity of use in agricultural landscapes (see Ponge et al, 2003; Federoff et al, 2005).

Below-ground biodiversity is sampled at plot level (sampling point or location). The grid defines a series of plots, with a target of about 100–120 plots per window. The plot is an area of at least 25×25m, determined by the minimum area needed to collect all required samples and carry out the required measurements, as described earlier. The principle of co-location is adopted for the sampling for both soil physical and chemical analyses and for the inventory of below-ground biodiversity. At the same time the windows allow for comparison of below-ground biodiversity between landscape components.

Location and definition of sampling window and grid

The windows are positioned within the benchmark areas in such a way that the major land uses are captured. Generally one window will not capture the major land uses in the benchmark area and separated smaller windows are therefore substituted. Figure 2.5 depicts the various configurations used. Figure 2.5a illustrates the use of one window with a grid of nine by eleven sampling points (plus some additional sampling points). Figure 2.5b depicts use of six, three and two separate windows. In the case where six separate windows are used, one window will measure 1km^2, depending on the distance between grid points.

In benchmark areas characterized by strong altitudinal gradients one expects the land use gradients to be oriented in the same direction, and the windows would then be located along that gradient (Figure 2.5b). At forest margins in the tropics it is often observed that a land use distribution pattern reflects a land use intensity gradient, with the more intense uses at greater distance from the forest. In such cases the sampling windows would be located along such a land use intensity gradient. Some knowledge and assessment of the region is therefore necessary before the selection of the sampling frames.

Where such gradients were not obvious, selection of windows aimed to cover two or three of the most important land uses and therefore also the most frequently observed

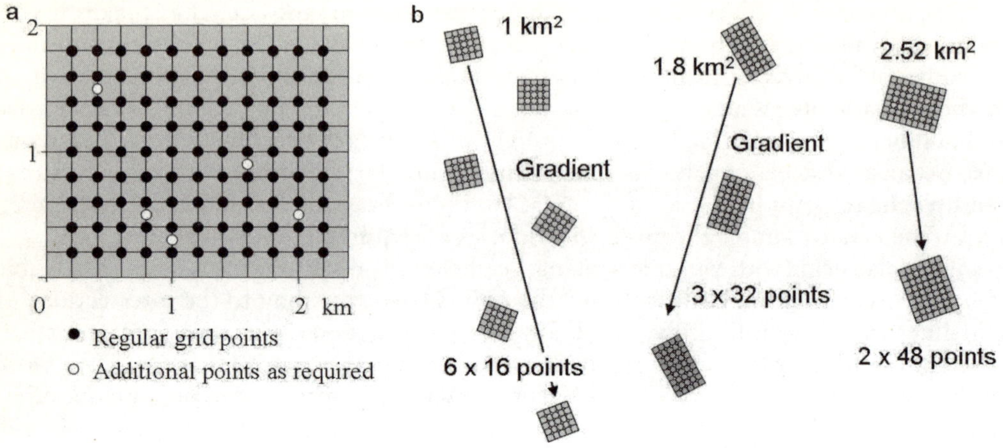

Figure 2.5 *Examples of different configurations of sampling windows: a) a full grid, with additional sampling points added as required; b) illustration of the division of a grid into six, three and two windows, positioned along a gradient*

transitions from one land use to another. For example, at forest margins this may be the transition from forest to grassland, forest to agricultural (crop) land, or forest to plantation cropping, with the different windows covering each typical transition in land use. This will allow for a comparative analysis of land use transitions (by relative measurements) while minimizing changes in below-ground biodiversity that arise from soil, topographic or climatic variations, rather than land use change per se. An element of replication is thereby introduced.

For the selection of the sampling windows aerial photos or high-resolution satellite imagery are used. Detailed land use and land-cover maps will also be suitable for this purpose. Where a digital land use map is available, the area occupied by each land use within any defined window can be determined using standard GIS software, which will then facilitate the selection of the windows. When only analogue materials are available, manual selection is made by visual interpretation, using a mask that can be moved over the photo.

Number of sample plots and grid size

The total number of sampling points per benchmark is fixed to be between 100 and 120. This can be obtained through various configurations of the sampling window(s) and the grid dimensions. Sampling points are added where the land use is highly variable. The total number of sampling plots is consistent with other sampling strategies proposed in similar biodiversity assessment exercises. For example, the BioAssess project (Ponge et al, 2003) recommends a similar number of sampling points, and the procedure recommended by the ASB project (Swift and Bignell, 2001) would amount to a comparable sampling effort. In the latter, each land use is assessed by at least one transect of 40×5m, from which at least five monoliths are dug for macrofauna sampling and five bulked samples for microsymbionts. Assuming six to seven distinct land use types per benchmark area (corresponding to the six to seven major land use classes generally defined in land use classification systems) with each land use sampled by three transects, this will result in 90 to 105 monoliths being sampled (and the same number of bulk samples for assessment of microsymbionts), approximately equal to the number of points in the grid sampling procedure described above. The main difference in the grid-sampling approach is that only one monolith and one bulk sample for microsymbionts are taken per plot. While sample sizes are much the same under transect-based and grid-based approaches, overall sampling effort is likely to vary.

In this connection Swift and Bignell (2001) argue, partly from logistical considerations, that to increase the reliability of data from their transect-based scheme, the number of sampling points should be increased from five to eight per transect (i.e. more points in the same plot), with an extended transect for some groups (for example, termites). In existing transect-based surveys, the ideal replication recommended by Swift and Bignell (three transects per land use) is almost never achieved, or else is assembled in a synoptic manner by combining transects carried out in different years or in unrelated projects (e.g. Eggleton et al, 2002; Davies et al, 2003).

Results from the inventory conducted by the CSM-BGBD project indicate that a minimum sampling effort of 100 to 120 points is required to capture most of the species within the area for at least some of the functional groups (e.g. termites, ants). With 100 to 120 points per benchmark area we generally reach the point where the species accumulation curve for macrofauna groups levels off. Capturing additional species would require considerable additional sampling effort however. Thus the sample design meets the objectives of inventory of the study area as far as assessment of species diversity is concerned. Smaller numbers or an uneven distribution of sampling points over the land use types generally results in a larger number of species not being captured and will make it more difficult to demonstrate differences in below-ground biodiversity between land use types that are generally characterized by a high variability in species richness.

The collection of samples is rarely the rate-limiting step in any biodiversity assessment exercise because the extraction, identification and cataloguing of specimens is by far the more time-consuming component of the process (Lawton et al, 1998). The experience of the CSM-BGBD project is that processing of 100 to 120 samples is feasible with the resources generally available. But even this requires a well-organized field campaign and established facilities for processing and analysis of the samples. If the capacity for handling and analysis of this quantity of samples is not available, the correct procedure to reduce the total number of sampling points would be to estimate, based on other data or based on a pilot study, the number of samples needed per land use to be representative of the below-ground biodiversity present, and then restrict the number of land uses investigated so that the total effort is compatible with the capacity for processing. Reducing the number of sampling points per land use category while maintaining the number of land use categories investigated may compromise the objective of producing statistically verified results.

Number of sampling points per land use category

With six to seven major land use types in each benchmark area, an average of about 15 points per major land-cover class results from grid-based sampling. With fewer major land use categories present, the number of points per land use category would increase.

The subsistence farming categories are prioritized to increase the number of samples, as most variation, due to management, history and so on, is expected to occur there.

Given the generally high level of variation in measurements of components of below-ground biodiversity, it is important to consider the minimum number of points required per land use category. The advice is to carry out a pilot to inform the sampling design. If the pilot confirms high variability in below-ground biodiversity within land use categories, tightening of the definition of the land use classes is performed. This implies adopting a higher level of specificity (detail) in the land use classification system, and subsequently selecting the land use classes that cover a clear gradient in land use intensity, according to the principles outlined in 'Dealing with variability' (above). The definition of a land use classification system is explained in Chapter 11.

Adaptation of the sampling approach to conditions in the field

Two strategies are available to assure appropriate representation of the major land use and cover classes in the sample. One strategy is to use a 'stratified grid' sampling approach as illustrated in Figure 2.6. For this purpose, a larger grid (enhancing the number of grid intersections) is defined such that the criterion for a minimum number of sampling plots is fulfilled for each of the land use types. Subsequently, a non-biased procedure is applied to reject a number of points to offset the increase in the total number sampling points that results from the enlarged grid, while satisfying the requirement for the minimum number of sampling points for each land use category. This procedure will also allow for the rejection of points that fall on non-targeted land use or land cover categories.

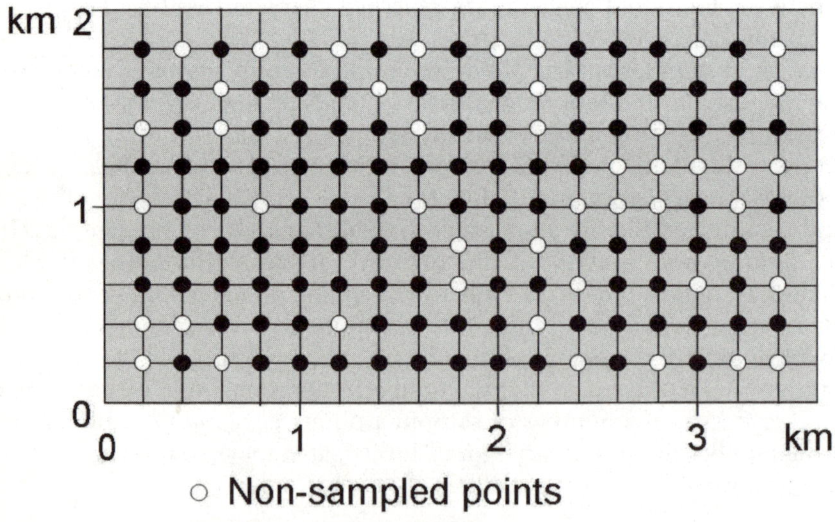

Figure 2.6 *Illustration of sampling points selected through a 'stratified grid' approach*

In the same manner one can start from an empty grid (only containing candidate sampling points) to randomly select sampling points for each land use category separately until the requirement for the minimum number of sampling points is fulfilled. This strategy implies allowing for irregular-shaped sampling windows. To select or reject sampling locations (plots) the land use for each grid point needs to be known, and requires land use to be inventoried prior to the actual sampling of below-ground biodiversity.

The second option is to select additional points within the window in between the existing grid points, for those land use categories that are under-sampled. This option is already illustrated in Figure 2.5a. The minimum distance between points (see below), needs to be respected. Although the insertion of additional points is clearly an expedient, some attempt should be made to randomize the selection of original points to be thus bisected.

The grid approach assumes general free access to the land where the sampling points are located. When this is generally not the case, an alternative sampling strategy needs to be adopted. For example, in the benchmark area in Mexico, access to farmers' fields was restricted. In this case fields were selected from farmers willing to allow access and a specified pattern for sampling within these fields was adopted, rather than a regular grid covering the larger area. In the Amazonian benchmark area in Brazil, due to the isolated location of communities, one to three windows were located in each available community area, though still adopting a fixed grid within those windows.

Distance between sampling points

Distributions of soil organisms may reflect landscape gradients, for example those of soil organic carbon, and cultivation practices. Distribution patterns of soil organisms may correlate with topography or vegetation patchiness or with soil resource gradients or even with the location of individual trees. On the other hand, distribution and grouping of soil organisms are also governed by intrinsic forces of population processes, such as dispersal, reproduction and competition (Lavelle and Spain, 2001). There are several theories to explain the extremely high species richness of soil communities and low degrees of resource specialization. In the case of the below-ground biodiversity project, with many different groups of soil organisms studied, the investigation of specific distribution patterns is not an intended objective, and sampling schemes that are suited for the analyses of spatial distribution patterns were therefore not considered. However, the grid sampling does allow some examination of spatial patterns at scales between the inter-sample distance and the grid size.

From the literature, examples can be found of soil biota spatially structured over distances of less than a centimetre to hundreds of metres at the field scale, depending on the organisms concerned. Ettema and Wardle (2002) mention that microbial biomass and Collembola are spatially dependent over ranges of more than 200 metres, that termites are spatially patterned over a range of up to 330 metres and nematodes' species distribution shows patch sizes ranging from 6 to 80 metres. Also recorded are patchy distributions at a miniature scale of centimetres to metres, nested within larger structures.

A 200m spacing seems acceptable because it allows a relatively larger area to be covered by the grid and is more likely to reflect the most dominant land uses in the sample, while traversing from one point to the other is still possible in most cases. At the same time it reduces the spatial dependence of observations for major groups of soil organisms such as earthworms. However, in the case of highly fragmented landscapes with very small fields (e.g. plots less than 0.5ha) and where traversing is severely hampered (e.g. very dense secondary regrowth in fallow plots) grid spacing can be reduced. For these reasons grid spacing was reduced in the Brazilian benchmark in the Amazon to 100m. Likewise if individual fields measure more than 2–4ha, increasing the distance between points should be considered.

Land use inventory and site characterization

The central hypothesis is that major variation in below-ground biodiversity is associated with difference in land use intensity described at the level of major land use categories. There are a number of interrelated factors, both biotic and abiotic (e.g. above-ground plant diversity, soil organic matter), and management characteristics (soil tillage, use of chemical inputs, etc.) that affect the distribution of soil organisms and it is assumed that the major land use categories provide for a meaningful distinction in terms of the factors determining below-ground biodiversity.

However, because both biotic and abiotic conditions as well as land use and land management characteristics may vary considerably between regions, soil biodiversity is inventoried along a land use intensity gradient for each of the benchmark areas separately. Concurrently information on annual rainfall, mean temperature and humidity, rain days, length of dry season, and cumulative seasonal rainfall up to the sampling date should be gathered for the window or even benchmark site in total, together with altitude, slope and aspect. Site history, with special reference to land use is also desirable (though not always obtainable).

Because little is known about the relative importance of the above-mentioned factors, the assumption that a land use category is at the right scale to accommodate their variation may prove to be wrong. It is possible that the variability in abundance and diversity of soil organisms within a land use category might be larger than variability between land use categories, especially in such diverse land use systems as typify the tropics.

Therefore, and because full site characteristics cannot be obtained from apparent land use alone, the approach is to include detailed description of land use and land cover, as well as of soil and site characteristics, as part of the inventory process, so as to allow for full data analyses to identify the parameters that explain the variability in below-ground biodiversity. The definition of land use classes therefore becomes important and is discussed in Chapter 11 of this handbook. The suite of basic physical and chemical soil properties should include bulk density, texture (S/S/C ratios), pF (soil moisture tension), pH, total C, total N, exchangeable cations, available P, CEC, Al^{3+} and H^+. Soil cores are taken from completely undisturbed ground but immediately adjacent to each monolith trench (the outer trench wall is probably the best place), thus providing the opportunity for correlating soil properties with the presence/absence of particular taxa and functional groups (see Anderson and Ingram, 1993).

Description of sites can be completed by the above-ground vegetation character. Species diversity and abundance of leguminous species are of particular importance in relation to the presence of Leguminosae-nodulating bacteria in the soil. Further, features such as mean canopy height, crown cover per cent, basal area, dominant cover/abundance scores for ground flora, litter accumulation and abundance, plant species and generic richness assist in arranging sites along botanical diversity gradients that have some relationship to their actual positions in the chronosequences and disturbance intensifications. Chapter 11 provides further detail on data requirements with respect to land use specifically. For sampling for other variables related to site characterization we refer to Anderson and Ingram (1993).

Sampling season

Soil organisms respond to seasonal changes, and results from the inventory of species richness and (relative) abundance will show marked differences between the seasons, depending on the methods used. With respect to arbuscular mycorrhizal fungi, for example, if the method relies on counting and identification of spores, the dry season might be the most appropriate for sampling. Spore production responds to plant phenology, and in case of annual crops spores are more abundant during the dry season due to root senescence, which may be different in, for example, forest areas. Spores are also best preserved during the dry season. In cases of phytopathogenic fungi and ectomycorrhizal fungi the evolution of structures and reproductive cycles (e.g. formation of sporocarps) are in synchrony with seasonal changes that occur within the host and are best observed during and towards the end of the rainy season. Also earthworm activity is highest towards the end of the rainy season (Tondoh and Lavelle, 2005). Ideally inventory should be carried out both during the wet and the dry season. However, in practice this is often not possible. For that reason sampling is carried out late during the rainy season with optional additional sampling during the dry season. The latter may be done with the specific aim of investigating seasonal changes in below-ground biodiversity.

References

Anderson, J. M. and Ingram, J. S. I. (1993) *Tropical Soil Biology and Fertility: A Handbook of Methods*, 2nd edition, CAB International, Wallingford

Cochran W. G. (1977) *Sampling Techniques*, 3rd edition, Wiley, New York

Davies, R. G., Eggleton, P., Jones, D. T., Gathorne-Hardy, F. and Hernandez, L. (2003) 'Evolution of termite functional diversity: Analysis and synthesis of local ecological and regional influences on local species richness', *Journal of Biogeography*, vol 30, pp847–877

Eggleton, P., Bignell, D. E., Hauser, S., Dibog, L., Norgrove, L. and Madong, B. (2002) 'Termite diversity across an anthropogenic disturbance gradient in the humid forest zone of West Africa', *Agriculture, Ecosystems and Environment*, vol 90, pp89–202

Ettema, C. and Wardle, D. A. (2002) 'Spatial soil ecology', *Trends in Ecology and Evolution*, vol 17, pp177–183

Federoff, E., Ponge, J-F., Dubs, F., Fernández-González, F. and Lavelle, P. (2005) 'Small-scale response of plant species to land-use intensification', *Agriculture, Ecosystems and Environment*, vol 105, pp283–290

Ford, E. D. (2000) *Scientific Method for Ecological Research*, Cambridge University Press, Cambridge

Gregoire, T. G. and Valentine, H. T. (2007) *Sampling Strategies for Natural Resources and the Environment*, Chapman and Hall/CRC Press, London

Kenkel, N. C., Juhfisz-Nagy, P. and Podani, J. (1989) 'On sampling procedures in population and community ecology', *Vegetatio*, vol 83, pp195–207

Lavelle, P. and Spain, A. V. (2001) *Soil Ecology*, Kluwer Academic Publishers, Dordrecht

Lawton, J. H., Bignell, D. E., Boulton, B., Bloemers, G. F., Eggleton, P., Hammond, P. M., Hodda, M., Holt, R. D., Larsen, T. B., Mawdsley, N. A., Stork, N. E., Srivastava, D. S. and Watt, A. D. (1998) 'Biodiversity inventories, indicator taxa and effects of habitat modification in tropical forest', *Nature (Lond)*, vol 391, pp72–76

Pedigo, L. P. and Buntin, G. D. (eds) (1993) *Handbook of Sampling Methods for Arthropods in Agriculture*, CRC, New York

Peterson, D. L. and Parker, T. (eds) (1998) *Ecological Scale: Theory and Applications*, Columbia University Press, New York

Ponge, J. F., Gillet, S., Dubs, F., Federoff, E., Haese, L., Sousa, J. P. and Lavelle, P. (2003) 'Collembolan communities as bioindicators of land use intensity', *Soil Biology & Biochemistry*, vol 35, pp813–826

Southwood, T. R. E. and Henderson, P. A. (2000) *Ecological Methods*, 3rd edition, Blackwell Science, Oxford

Stein, A. and Ettema, C. (2003) 'An overview of spatial sampling procedures and experimental design of spatial studies for ecosystem comparison', *Agriculture, Ecosystems and Environment*, vol 94, pp31–47

Swift, M. and Bignell, D. E. (eds) (2001) *Standard Methods for the Assessment of Soil Biodiversity and Land-use Practice*, International Centre for Research in Agroforestry, South East Asian Regional Research Programme, ASB-Lecture Note 6B, Bogor, Indonesia, www.worldagroforestrycentre.org/SEA/Publications

Thompson, S. K. and Seber, G. A. F. (1996) *Adaptive Sampling*, Wiley, New York

Tondoh, E. J. and Lavelle, P. (2005) 'Population dynamics of *Hyperiodrilus africanus* (Oligochaeta, Eudrilidae) in Ivory Coast', *Journal of Tropical Ecology*, vol 21, pp493–500

Urban, D., Goslee, S., Pierce, K. and Lookingbill, T. (2002) 'Extending community ecology to landscapes', *Ecoscience*, vol 9, pp200–202

Webster, R. and Oliver, M. A. (1990) *Statistical Methods in Soil and Land Resource Survey*, Oxford University Press, Oxford

Wessels, K. J., van Jaarsveld, A. S., Grimbeek, J. D. and van der Linde, M. J. (1998) 'An evaluation of the gradsect biological survey method', *Biodiversity Conservation*, vol 7, pp1093–1121

Chapter 3

Macrofauna

David E. Bignell, Reginaldo Constantino, Csaba Csuzdi, Agus Karyanto, Suleymane Konaté, Julio Louzada, F.-X. Susilo, Jérôme Ebagnerin Tondoh and Ronald Zanetti

Introduction

In this chapter we deal with methods for the inventory of soil macrofauna. The macrofauna group comprises those soil animals that are more than 1cm long or have a width or diameter of more than 2mm. A diversity of soil organisms is included in this category (Chapter 1), but in this chapter we concentrate on the most significant groups: the earthworms, termites, ants and beetles. The earthworms are probably the most important soil invertebrates in temperate regions, but in tropical environments this role is taken over by termites and ants. Many macrofauna play an important role in soil ecosystems as soil engineers, especially in tropical environments, significantly influencing the physical and chemical properties of the soils they inhabit, especially by the creation of macropores and transformation and redistribution of organic matter. Others are comminuters, shredding organic matter, and a number of groups are macropredators. Because of their important role in ecosystem processes and their sensitivity to the environmental conditions, macrofaunal groups are often used (or proposed) as indicators of soil biological quality and therefore form an important component of the soil biota, indicative of overall soil biodiversity and effects of land use change and management practices. Further, because of their generally high relative biomass they form an important constituent of the foodweb in the soil. For the use of abundance biomass and diversity data as indicators for monitoring effects of land use change, standardized methods need to be applied. This chapter describes proposed methods for the inventory of macrofauna as they were applied in the CSM-BGBD project. The proposed methods are intended for application in tropical environments, but may nevertheless be more widely applicable.

In Chapter 2 it is explained that there is no single method for the inventory of a specific group of macrofauna. Generally, some representatives of the target macrofaunal groups are captured with any of the methods listed in Chapter 2. Therefore, this chapter is structured according to the method rather than according to functional group. Nevertheless each of the methods described is particularly suited for one particular group: the monoliths are especially suited for the inventory of earthworms, the transect method is particularly suited for the inventory of termites, the Winkler extraction is the most suitable for capture and extraction of ants from litter samples whereas the (baited) pitfall traps are suited for capture of beetles. Therefore, in each of the following sections describing a particular method, the emphasis is on the associated target group of organisms as well the generality of macrofauna.

As with several other groups of the soil biota, the starting point for the development of our protocols has been the procedures of Anderson and Ingram (1989, 1993), consolidated and improved by Swift and Bignell (2001) and which are generally referred to as the 'TSBF methods'. These were based on soil monoliths and a short transect (20/40×2m) for termites, to which we have now added pitfall traps (principally for beetles) and some guidelines for casual sampling. A major advance is the proposal to use the transect principle for ants and beetles, extracting the insects from litter quadrats by the Winkler method (Bestelmeyer et al, 2000). New schemes are also proposed for the arrangement of monoliths to improve the sampling of earthworms. The fact that we can devote a whole chapter to macrofaunal sampling, as compared to the three printed pages in the original 'TSBF methods' demonstrates how our perception of the importance of these organisms has grown over the intervening years.

A. Inventory of soil and litter macroarthropods and earthworms in particular using the soil monolith

Overall soil and litter macroarthropods – TSBF monolith

General procedures follow Anderson and Ingram (1993), with the addition of pitfall traps (Swift and Bignell, 2001) and Winkler bag extractions (Bestelmeyer et al, 2000), together with the digging of additional soil monoliths of dimensions 50×50×20cm (depth) for earthworms. For discussion of the number and positioning of monoliths see Chapter 2. An additional 20×2m sampling transect is used for termites. This transect can also be used for ants and beetles, although it is better if litter is sampled from quadrats arranged in a separate parallel line.

A quadrat of 25×25cm is first put down to mark the position of the monolith, and any leaf or woody litter within it removed and bagged (Figures 3.1a and 3.1b). Isolate the monolith by cutting down with a spade a few centimetres outside the quadrat and then digging a 25cm (wide) and 30cm (deep) trench around it, isolating the monolith as an undisturbed pillar (Figures 3.1c and 3.1d). In a variant of the method, all invertebrates longer than 10cm excavated from the trench and the litter can be collected; these will be mainly large millipedes and earthworms with low population densities but representing an important biomass. Their abundance and biomass can be calculated on the basis of 0.50m^2 samples (i.e. the approximate width of the monolith plus two trench widths squared). Divide the delimited monolith block into three layers, 0–10cm, 10–20cm and 20–30cm. This can be done conveniently using a machete or parang held horizontally and grasped at both ends. Hand sort the litter and each monolith layer separately, preferably on site and under shade. If time is short or the light poor (sorting in closed canopy forest is usually difficult as light fades in the late afternoon), bag the soil and remove to a laboratory. Ants can be extracted by gently brushing small (handful) quantities of soil through a coarse (5mm) sieve into a tray: the sieve retains the larger particles, while the ants can be pootered up from the tray. Bagged soil should be kept out of direct sunlight and sorted within 24 hours (but preferably sooner). For soil, it is often convenient to use the surface of a large field-portable table for sorting so that several assistants can work on the same sample, or different samples, simultaneously. For litter, a large plastic kitchen tray with an upraised margin makes a good sorting arena. Woody items should be dissected into small pieces to discover insinuated termites, ants and beetles. Small twigs should be broken and the pieces tapped firmly on the tray to dislodge any insects they contain. For labelling, the ants, termites and beetles collected should be preserved in alcohol; for termites include soldiers as well as the worker caste. Earthworms and molluscs are generally preserved in formalin to reduce the exudation of mucus. One specimen tube should be used for each population (or apparent colony) encountered. A label should be added into the alcohol on which is noted (NB: in pencil or waterproof ink, not ballpoint):

Note: Upper illustrations show the placement of a wooden marker frame to locate the position of the monolith and the clearance of loose litter into a sorting tray. Lower illustrations show the large (50×50×30cm) and small monoliths and the excavation of the surrounding trench before the monolith itself is sampled.

Figure 3.1 *Marking and excavation of a small monolith (25×25×30cm)*

- date and site identification number;
- for transects or aligned monoliths, the section number along the transect where the specimen occurred (1, 2, 3 or 4);
- the microhabitat concerned (new dead wood, rotten wood, mound, soil, litter, soil at tree root, etc.).

The information is important in establishing the nature of the organisms discovered (especially functional group diversity) and for constructing a species accumulation curve. To avoid confusion in rich sites, the labels should be written as soon as the specimens have been placed in the tubes. Elsewhere, labels can be written when the searching of each section of the transect is completed. However, when searching is timed, as in each section of a transect, the period allocated should include labelling time.

An alternative sampling protocol for assessment of earthworm abundance and diversity in tropical ecosystems

Introduction

This alternative protocol combines the so-called TSBF soil monolith (25×25×30cm) and three large soil monoliths (50×50×20cm) to assess the abundance and diversity of earthworms at local and landscape levels in tropical ecosystems. It satisfies the requirement of rapidity and the need to take soil heterogeneity at plot level into account.

For over a decade, most field studies on earthworms have been undertaken using the TSBF method (Anderson and Ingram, 1993) for soil macro-invertebrates; see Gilot et al (1995), Haynes et al (2003), Dlamini and Haynes (2004) for sampling of earthworms in Africa; see Bhadauria et al (2000) for Asia; and see Decaëns et al (1994), Barros et al (2002), Decaëns et al (2004), Mathieu et al (2004) for examples from South America. In all cases the unit of sampling is a square soil monolith of 25×25cm and 30cm depth, the monoliths being distributed along a short (40m) transect, with at least five monoliths per study plot.

Despite the relative success of the TSBF method to estimate soil macro-invertebrate abundance and diversity, Eggleton et al (1999) and Agosti et al (2000) have proposed more exhaustive sampling protocols for termites and ants (respectively) because of their functional group specificity, the relative scarcity of colony centres and their highly patchy distribution. Similar arguments can be used to justify more extensive sampling of earthworms. Although small soil monoliths have been successfully used to show significant changes in earthworm diversity at plot level under different land management practices (Fragoso et al, 1997; Decaëns and Jiménez, 2002), few investigations have focused on the assessment of earthworm abundance and diversity in agricultural landscapes.

Originally, the demographics of earthworm populations were studied using large soil monoliths of 1m^2 (100×100×40cm depth, e.g. Vincent, 1969; Lavelle, 1978), but they are time-consuming and tedious. Recently, the more efficient combination of large (100×100×40cm) and small (25×25×30cm) soil monoliths has been reported to survey population dynamics of the peregrine earthworm *Hyperiodrilus africanus* (Tondoh and Lavelle, 2005). Medium-sized (50×50×40cm) monoliths were used by Ortiz-Ceballos and Fragoso (2004) to assess the impact of mulching on earthworm populations. Since contemporary agricultural practices result in a mosaic of land use types covering large

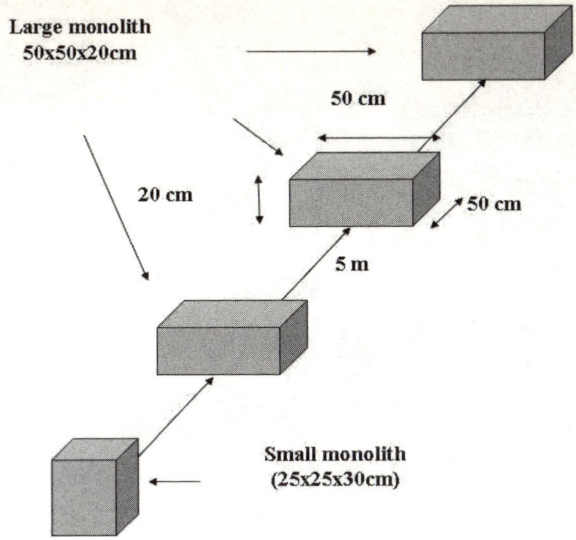

Figure 3.2 *Scheme for the arrangement of supplementary earthworm monoliths*

areas, the challenge is to develop rapid sampling protocols for sampling at the landscape level. Such an advance is crucial for developing indicators of land degradation (Burel et al, 1998; Moreno and Halffter, 2000, 2001), as earthworms are known to be sensitive to anthropogenic disturbance (Fragoso et al, 1997; Fragoso et al, 1999; Decaëns and Jiménez, 2002).

Sampling protocol

Sampling should ideally be conducted at the end of the rainy season when earthworms are known to be more active (Tondoh and Lavelle, 2005). At each sampling point, one small monolith (25×25×30cm) is dug out and stratified into three layers: 0–10cm, 10–20cm and 20–30cm. Earthworms are collected by hand sorting.

Five sampling points are randomly chosen in each land use type. This stratified sampling scheme, which combines smaller and larger monoliths, allows for plot level variation in populations to be taken into account, while the variation at landscape level is addressed by sampling of the different representative major land uses.

At each selected point, earthworms are sampled following TSBF sampling procedure, as above (Anderson and Ingram, 1993). In addition, three large monoliths of 50×50cm and 20cm deep are dug out at 5m intervals along a transect with a randomly positioned starting point but perpendicular to the slope if appropriate (modified TSBF protocol, Anderson and Ingram, 1993). The transect is schematically represented in Figure 3.2. The additional monoliths are excavated in order to take into account heterogeneity at plot level. Each larger monolith is separated into two layers (0–10 and 10–20cm) rather than three layers because most of the individuals (57 to 99 per cent) are located in the top 20cm during the rainy season (Tondoh and Lavelle, 2005). Earthworms are hand sorted (Figure 3.3) and stored in 4 per cent formalin solution for identification. Individuals are then identified, counted and weighed.

Figure 3.3 *Hand sorting of earthworms in plastic trays*

Earthworm identification and estimations of abundance

Earthworm specimens, previously sorted out according to morphospecies, can be identified to family level using the keys developed by Blakemore (2005). The common peregrine earthworm species distributed throughout the tropical region are identifiable to species level using the keys and descriptions provided by Blakemore (2002). However, the native species require country-specific expertise due to lack of comprehensive identification keys.

For the African Eudrilidae and Acanthodrilidae (Benhamiinae), the revisions of Sims (1980, 1987) and Csuzdi (1996) can be used and where it is necessary the original species descriptions need to be consulted. In the identification of the Indian and Southeast Asian earthworms, the monographs of Gates (1972), Sims and Easton (1972), Easton (1979) and Julka (1988) are appropriate. Unfortunately, there are no up-to-date monographs available for identification of Central and South American worms. Therefore the review of Eisen (1900) and selected papers from several authors (Righi, 1971, 1995; Fragoso and Rojas-Fernandes, 1994; Zicsi, 2001, 2005) are the main points of reference. Earthworm abundance is estimated as the number of individuals per m^2 and biomass as g of individuals per m^2. Earthworm diversity can be studied at two complementary levels: taxonomic and functional diversity (Bouché, 1977; Lavelle, 1983; Fragoso et al, 1997). The functional aspect of diversity is of greater interest for assessment of the relationship between diversity and ecosystem services.

The first level of taxonomic diversity deals with the number and the identity of earthworm species. Three expressions of species richness can be used: (1) the number of species recorded in all monoliths per land use type, (2) the average diversity expressed as the mean of species number per land use type and (3) the Shannon–Wiener index of diversity (Pielou, 1996). The second level of taxonomic diversity concerns earthworm biogeography (Fragoso et al, 1997). In a given ecosystem, two groups of earthworms can be distinguished according to their origin: native or exotic species. Native species are characterized by a restricted distribution at local, regional and continental levels. Exotic species are in most cases introduced through human interference. They have been called peregrine to point out their wide distribution at regional and continental levels (Lee, 1987). The ratio of native to exotic can be used as an index for native species extinction or as an invasion index of exotic or peregrine species. The analysis of earthworm communities through this approach is done by identifying in each land use type native and exotic or peregrine species. The structure of earthworm communities can be characterized through the density and the biomass of native and exotic worms.

The importance of functional diversity of earthworms has been widely documented (Lavelle, 1983; Blanchart et al, 1997; Fragoso et al, 1997; Lavelle et al, 1997). Ecosystem services provided by earthworms are related to the impact of their activity on the soil system (Lavelle et al, 1997). From an ecological point of view, earthworm communities can be divided into three groups (Bouché, 1977; Lavelle, 1981):

1. **Epigeic** earthworms: small-sized with pigmentation and living in the litter system at the soil surface.
2. **Endogeic** earthworms: medium-size worms, unpigmented and living within the soil. As a result of their activity, casts are produced within the soil profile and at the soil surface.
3. **Anecic** earthworms: they are large earthworms, and unpigmented. They live in soil but feed at the surface. Their activities result in dense network of burrows within the soil.

Most earthworms belonging to the endogeic and anecic type are denoted ecosystem engineers (Lavelle et al, 1997) because they are active in mixing and digging the soil, modifying the hydraulic and chemical properties. Earthworms can be assigned an ecological category or functional group and the groups compared on the basis of their density and biomass.

B. Transect methods for termites, ants, beetles and earthworms

Introduction

This method, which uses a transect of 100, 50 or 20×2m, was developed for forests or sites recently derived from forests (as in slash-and-burn agriculture) and follows the formal descriptions by Jones and Eggleton (2000) and Swift and Bignell (2001). The data obtained are qualitative (species identity and number) and/or semi-quantitative (relative abundance of species and/or specified functional groups), but are considered to have relatively high resolution, and are therefore treated separately from the monoliths and pitfall traps. The method is low-impact, is suitable for non-termite specialists and can be completed (at each sampling point) by two people in a day (two days for a 100m transect). The method, as developed and as described below, was based on a 100m transect, but this has been abridged to 20m in the standard protocol for CSM-BGBD and to 50m in the alternative sampling scheme (see Chapter 2).

Equipment required

Compass, 30m tape measure, or (preferably) nylon rope or string marked into 5m sections using yellow or orange fluorescent ribbon or waterproof tape, 2m pole. For each person, a sharpened machette or parang, trowel, high-sided plastic or metal tray, two pairs of forceps, ca. 40 stoppered vials (approx 1×5cm) containing 80 per cent alcohol (more vials in termite-rich habitats). A short fixed-bladed camping knife (ca. 8–10cm) may also be useful for probing wood and soil.

Procedure

Lay out a transect line of 20, 50 or 100m adjacent to the monolith, though at sufficient distance to avoid any mutual disturbance during sampling. The transect should be placed to run through visually homogeneous surroundings, avoiding such features as large streams, cliff edges or fresh skid trails which are not suitable habitats for termites. However, the transect can incorporate other natural features of the biotic environment which contribute to its physical heterogeneity, such as hill slopes, narrow stream gullies or small canopy gaps. Some subjective judgement is often necessary to decide on the most suitable line, especially where the treatment plots concerned are small. Transect lines do not need to be absolutely linear, and can be turned through angles of up to 90° to avoid natural obstacles, as long as they do not re-intersect with themselves. In small plots a transect can be turned, successively, through two 90° angles to run back towards the starting point, but the two main 'arms' of the transect should be at least 15m distant. Alternatively, two 50m (or 10m) transects can be run in parallel. If the line runs through the middle of a tree with a very large basal area, the transect can be bent to one side at that point, as long as at least a part of the rooting system falls within the 2m-wide sampling belt. Make a note of the starting point, initial compass bearing of the transect line and any major directional changes. If other transects are employed at the same site (for a replicated monolith, for example), transects should be positioned so as to avoid mutual interference and separated to reduce excessive trampling in one place. Keep pitfall lines undisturbed at all times.

The time required for the 20m transects is 30 minutes of sampling by two people in each section of 5m. A team of two people can sample 8–12 sections a day (i.e. 2–3 transects). The time can be reduced by working in multiple teams of two people simultaneously in different sections. Sampling is stratified by niche. The following microniches are investigated in detail: surface soil and litter down to a depth of about 5cm; deep accumulations of litter and soil between large buttress roots; dead wood in all stages of decay; termite carton or soil runways on tree trunks and other vegetation; subterranean, epigeal and arboreal termite nests and mounds to a height of 2m above ground level (including purse nests suspended on vegetation). Soil, litter and woody items can be rapidly dissected in trays; this is often helpful in revealing termites (Figure 3.4), beetles and earthworms in cryptic niches, or when light is poor. Sticks should be broken into pieces and tapped onto the trays with moderate force to displace any termites and ants they contain. Larger items of dead wood should be chopped up, bearing in mind that they may be infested in one part and not another. Rotting wood partly incorporated into the topsoil or covered in soil sheeting will frequently contain termites. Mounds and nests may be inhabited by species other than, or in addition to those building them; it is therefore advisable to check the periphery and the base of the

Note: Fine forceps are suitable for picking the insects up undamaged, if care is taken. Alternatively a fine artist's brush moistened with alcohol can be used. Note the collecting tube ready in the left hand of the assistant.

Figure 3.4 *Sorting termites from a trowel sample of soil in an aluminium tray*

structure, as well as its central chambers. The trunks of living trees can be investigated for termite runways as far up as they are easily accessible without ladders or climbing (about 2m), not forgetting that forks or tree ferns can accumulate soil which contains termites, earthworms and possibly other invertebrates.

A short training or orientation period is usually necessary before inexperienced collectors can sample with the same efficiency as experts. Ideally, a training transect of 50 or 100m should be sampled first, with the guidance of an experienced collector. Collectors should work steadily (rather than frantically) through each 30-minute sampling period and aim to maintain the same level of sampling efficiency in all sections of the transect. For this reason, and to minimize the necessity of having to work in poor light, it is recommended that no more than 12 sections be completed in any one day. It is also helpful to rest for a few minutes between sections. In most sites there is no need to collect every termite found, and very common species might be passed over after being initially sampled in order to search for rarer or more cryptic forms, or to find soldiers in species that have relatively low soldier/worker ratios (but bearing in mind that some species are soldierless). It is impossible to sample efficiently in heavy rain, so it is permissible to interrupt collection until better conditions return. Work can be divided between the collectors in any mutually agreed way, for example for each person to work in the 1m belt on opposite sides of the line, to divide each section into two subsections each of 2.5×2m, for one to sample wood and mounds, while the other examines soil and tree roots, etc., depending on the nature of the site and the section-to-section topography. It is recommended that soil should be dug up in at least a dozen well separated places per section of transect. In transects where few termites are encountered, it is important to observe the sampling protocol exactly, in spite of the small returns, and not to curtail the work.

Processing of specimens

Representative specimens of the termites discovered should be preserved in alcohol, wherever possible including soldiers as well as the worker caste. One specimen tube should be used for each population (or apparent colony) encountered. A label should be added into the alcohol on which is noted (NB: in pencil or waterproof ink, not ballpoint):

- date and transect identification number;
- the section number along the transect where the specimen occurred (1, 2, 3, etc.);
- the microhabitat concerned (new dead wood, rotten wood, mound, soil, litter, soil at tree root, etc.).

Alternatively, print sequentially numbered field labels in a laser printer, cut and stick them to a field notebook in sequential order. In the field just add the labels to the vials and take notes in the field notebook on the range of numbers per section and transect. This makes the fieldwork easier and all labels follow the same pattern and are easier to read, also reducing the chances of numbering errors. All field notes are kept in the notebook, and therefore linked to the label numbers. The information on the microniche where each specimen is found is important in establishing the nature of the termite community (especially functional group diversity) and for constructing a species accumulation curve. To avoid confusion in termite-rich sites, the labels should be written (or field notes written) as soon as the termites have been placed in the specimen tubes. Elsewhere, labels or notes can be written when the searching of each section of the transect is completed. However, the 30 minutes allocated should include labelling time.

Specimens can be damaged and their colour affected by soil and debris in the vial. Thus, specimen tubes must be cleaned as soon as possible (i.e. the termites separated from soil and other debris, fresh alcohol added and new neat labels written) on the same day as the collection takes place, or as soon as possible afterwards. This speeds subsequent processing of the material by expert taxonomists, and is another reason why a minimum of two days should be allowed for each 100m transect attempted (pro rata for shorter transects). The termites should be separated into recognizable taxonomic units, either morphospecies or named species. By treating each 5m section as an independent sample, a specimen-based species or species accumulation curve can be constructed. Ten random sequences of sections are generated by drawing (in each sequence) 20 sections at random without replacement. The number of species found in each section is then used to calculate the cumulative number of species for each selection of the ten sequences. Finally, the mean cumulative number of species of the ten sets of 20 sections is calculated for each section and a species accumulation curve drawn. An ideal curve should be asymptotic (i.e. should rise, then flatten out) indicating that few new species would be found by additional transect sampling in the area. The species richness of the site can be estimated by extrapolation methods such as first-order Jackknife, second-order Jackknife, Chao and ACE, among others (Magurran, 2004), which are implemented in the free software EstimateS (Colwell, 2000, 2005). The numbers of species and morphospecies encountered per transect can range from just a few to more than 50, depending on the site and the biogeographical region.

Identification

The first step in producing a species list and organizing the data on the species assemblage is the identification of the species present in the sample. Identification is the determination of the position of a specimen in the existing classification, working from higher levels, such as orders, down to families, genera and species. Ideally all specimens should be identified to species, but this is often not possible for some groups of soil organisms. More realistically, the final list will be partially identified to species and the rest will be identified to genus and sorted into morphospecies. When comparing data from different habitats or localities, it is important to use the same identification effort and methods. It also important that morphospecies 'A' of site '1' is the same as morphospecies 'A' in site '2'. Poor taxonomy may result in underestimated or overestimated species diversity and similarity and lead to erroneous conclusions. Identification of soil organisms to species usually requires high taxonomic expertise and is a very time-consuming process. Samples and specimens need to be carefully labelled and organized. The whole collection should be preserved for some time after the identification for confirmation and a representative subset needs to be deposited in a permanent national institutional collection or museum.

Common methods of identification of soil fauna include a) use of published identification keys, b) comparison with descriptions, c) comparison with reference collections and d) sending specimens to an expert taxonomist. Molecular techniques are routinely used for the identification of microorganisms, but currently not for animals. The format of the names must follow the respective code of nomenclature. For animals this is the International Code of Zoological Nomenclature (International Commission on Zoological Nomenclature, 1999); for plants, the International Code of Botanical Nomenclature (Greuter et al, 2000); and for bacteria, the International Code of Nomenclature of Bacteria (Sneath, 1992). The author and year of publication are not part of a species name and their use is optional, although often recommended, particularly in diverse or poorly known groups of organisms. However, the codes of nomenclature have strict rules of format for author names, which must be followed. The codes of nomenclature recommend that the names of genera and species be printed in italics, but this is not mandatory. Never use manuscript (unpublished) names. Consult a taxonomic catalogue to check the validity and correct spelling of the names if one is available.

Table 3.1 presents an example of a species list. The list is organized in alphabetical order, which is usually more practical than some subjective 'taxonomic order'. Three different levels of identification are shown. For example, *N. corniger* and *N. similis* were fully identified to species. *Termes* sp. A and *Termes* sp. B were identified only to genus and morphospecies. The abbreviation 'cf.' in *Coatitermes* and *Armitermes* (from the Latin *confer*, compare) is used to indicate that the identification is uncertain. Other abbreviations are sometimes used with the same or similar meaning. Note that the abbreviations 'sp.' and 'cf.' are not in italics. The parentheses around the names of authors indicate that the species was originally described in another genus and later transferred to the current one. The parentheses, therefore, carry a meaning and their use is determined by the rules of nomenclature.

Table 3.1 *Example of species list*

No	*Species*	*forest*	*fallow*	*crops*	*agroforestry*	*pasture*
1	Kalotermitidae			1		
2	*Angularitermes nasutissimus*			1		
3	*Angularitermes* sp.		1			
4	*Anoplotermes* sp. 1	6	13			4
5	*Anoplotermes* sp. 2	1	9	2	3	10
6	*Anoplotermes* sp. 3	6	13	19	1	71
7	*Anoplotermes* sp. 4	25	46	20	9	47
8	*Anoplotermes* sp. 5	8	2			3
9	*Anoplotermes* sp. 6	1				
10	*Anoplotermes* sp. 7		2			
11	*Anoplotermes* sp. 8	1	1			
12	*Anoplotermes* sp. 9		15	10	1	
13	*Anoplotermes* sp. 10					2
14	*Anoplotermes* sp. 11	7	7			16
15	*Anoplotermes* sp. 12		1			
16	*Anoplotermes* sp. 13			1		
17	*Anoplotermes* sp. 14	1				1
18	*Anoplotermes* sp. 15		2	1		
19	*Anoplotermes* sp. 16	1	6	7	4	
20	*Anoplotermes* sp. 17	1		3		14
21	*Anoplotermes* sp. 18	2				
22	*Anoplotermes* sp. 19			4		
23	*Araujotermes nanus*		1			1
24	*Armitermes holmgreni*		3			
25	*Armitermes minutus*	1				
26	*Armitermes peruanus*	3	1	1	5	
27	*Armitermes teevani*				1	
28	*Atlantitermes osborni*	1	2			
29	*Caetetermes taquarussu*	1	1			
30	*Constrictotermes cavifrons*	1				
31	*Coptotermes testaceus*	5	10	8	3	2
32	*Cornicapritermes mucronatus*	1	1			
33	*Cornitermes pugnax*	10	22	7	4	
34	*Cornitermes* sp.	9	4			
35	*Curvitermes odontognathus*		1			1
36	*Cylindrotermes flangiatus*	10	1			
37	*Cylindrotermes parvignathus*	7	1			
38	*Embiratermes parvirostris*	1	3	1		
39	*Heterotermes tenuis*	18	85	56	29	6
40	*Labiotermes psilus*	2				
41	*Microcerotermes strunckii*	2	1			
42	*Nasutitermes callimorphus*	3	6	2	2	1
43	*Nasutitermes corniger*	15	29	11	16	1
44	*Nasutitermes gaigei*	1	1	1		
45	*Nasutitermes guayanae*	12	35	9	1	
46	*Nasutitermes octopilis*		6	1		
47	*Nasutitermes* sp. A	17	1	5		
48	*Nasutitermes* sp. I	21	89	29	37	19

No	Species	forest	fallow	crops	agroforestry	pasture
49	*Nasutitermes* sp. J		4	1		
50	*Nasutitermes surinamensis*		2	1		1
51	*Nasutitermes unduliceps*	1				
52	*Nasutitermes wheeleri*		1			
53	*Neocapritermes pumilis*	2				
54	*Neocapritermes* sp.	6		1		
55	*Neocapritermes talpa*	1	1			
56	*Neocapritermes taracua*		1	2		
57	*Neocapritermes villosus*	6		3		
58	*Obtusitermes* sp.	8				
59	*Planicapritermes planiceps*	1				
60	*Rhinotermes hispidus*	1	9	6		
61	*Rhinotermes marginalis*	1	3	5		
62	*Rotunditermes bragantinus*	2				
63	*Rugitermes* sp.	1				
64	*Ruptitermes* sp. 1	1				
65	*Ruptitermes* sp. 2	1				
66	*Ruptitermes* sp. 3		1			
67	*Subulitermes baileyi*			1		
68	*Subulitermes microsoma*		1			
69	*Syntermes molestus*	5	5	4		
70	*Syntermes spinosus*	1				
71	*Syntermes territus*	1	2			
72	*Termes hispaniolae*			2		
73	*Termes medioculatus*	1	1			3
74	*Triangularitermes* sp.		3			
75	*Triangularitermes triangulariceps*	9	8	2		
76	*Velocitermes* sp.	9	7	1	3	
	Samples	259	471	229	119	203
	Species	52	50	35	15	18
	Transects	20	39	18	10	13
	Shannon Index (H)	3.41	2.91	2.80	2.04	2.01
	Simpson Index	0.96	0.90	0.90	0.81	0.80
	Rarefaction (100 samples)	34.4	26.6	25.3	14.3	14.1
	Chao estimate	86.7	81.8	52.3	17.2	25.3
	Jackknife 1 estimate	74.8	72.4	50.1	18.6	25.4

List of identification guides and catalogues

An integrated list of identification guides for termites, ants and beetles is given in Appendix 1, with more specialized lists of taxonomic publications for termites in Appendices 2 and 3. Bibliographic details are given in the literature list at the end of the chapter, if not given in the Appendices itself. Inevitably, the availability of suitable published keys and their quality varies between taxa and between regions, and many museums or other scientific institutions with an interest in taxonomy and biodiversity conservation have their own internal manuals, which are not widely available outside. Also, there is an extensive 'grey literature' of workshop reports, conference proceedings and documents produced by international development projects which do not have global distribution, or focus on one region only. The best approach to identification is therefore to make arrangements for training, or for the processing of specimens, with an established taxonomist before sampling is undertaken.

Interpretation of results

Species assemblages are groups of taxonomically related species from a specific habitat or geographic area. They are sometimes referred to as 'species communities', but this term is usually not correct because such assemblages are delimited by taxonomic classification, which is based on history and phylogeny and not on current ecological relations. The assemblages may or may not form a uniform functional group. The composition of the assemblage is also often affected by the sampling methods used. The characterization of a species assemblage involves several steps, from field sampling to identification and data analysis. One of the main properties of an assemblage is 'species diversity', which may be defined and estimated by various methods. Most often, the sample will not contain all species present in the habitat. Therefore, it is necessary to extrapolate from the sample and use comparable estimates of local diversity. Assemblages from different habitats or localities can be compared with various estimates and indices, and can be grouped based on their similarities. Although the basic information generated is species richness, other information can be obtained from the transect. Most important is functional group diversity; this can be obtained by noting the numbers of species and morphospecies. Termites can be placed in the following trophic categories:

- *Soil-feeders.* Termites distributed in the soil profile, the organic litter layer and/or epigeal mounds, feeding deliberately on mineral soil, apparently with some degree of selection of silt and clay fractions. Although the ingested material is highly heterogeneous, there are higher proportions of soil organic matter and silica, and lower proportions of recognizable plant tissue than in other groups (Sleaford et al, 1996).
- *Soil/wood interface-feeders.* Termites feeding in highly decayed wood that has become friable and soil-like, or predominantly within soil under logs or soil plastered on the surface of rotting logs or mixed with rotting leaves in stilt-root complexes. This group is synonymous with 'intermediate feeders', sensu DeSouza and Brown (1994), but not the same as the category 'rotten wood-feeders' recognized by Collins (1989).
- *Wood-feeders.* Termites feeding on wood and excavating galleries in larger items of woody litter, which may become colony centres. This group also includes termites having arboreal, subterranean or epigeal nests but feeding elsewhere, and many Macrotermitinae cultivating fungus gardens. 'Wood' includes dead branches still attached to living trees and dead standing trees as well as fallen larger items that are fresh, or in all except the terminal stages of decay.
- *Litter-foragers* (or harvester termites). Termites that forage for leaf litter, live or dry standing grass stems and small woody items, usually cutting the material before consumption or carrying it to the nest system. This includes some subterranean and mound-building Macrotermitinae, as well as certain Nasutitermitinae that forage on the surface of the ground, and at least one lower termite in southern Africa, *Hodotermes mossambicus*, with a similar habit. Foraging termites are usually more conspicuous than other feeding types, by reason of the numerous galleries or soil sheets constructed over wood, litter and the surface of the ground, foraging holes

opening to the surface from subterranean passages or the formation of uncovered columns of individuals (usually mixtures of foraging workers and soldiers).

- *Specialized- and incidental-feeders*. This category follows the listing of termite foods given by Wood (1978), and includes species feeding on fungi, algae, bryophytes and lichens on tree bark (for example *Hospitalitermes hospitalis* in Southeast Asia and some species of *Constrictotermes* in South America feed on lichens), dung-feeders and scavengers of vertebrates' corpses (probably consumed opportunistically, although dung is arguably a form of decayed litter), and also certain secondary inhabitants of termite mounds which feed on the organic-rich linings of internal chambers as obligate inquilines (for example *Ahamitermes* and *Incolitermes* in Australia, Gay and Calaby, 1970; *Ophiotermes* and *Tuberculitermes* in West/Central Africa, Eggleton and Bignell, 1997; *Inquilinitermes* living inside nests of *Constrictotermes*). Some termites also feed on living plant tissue, although no termite species is really phytophagous.

The categories are not mutually exclusive and many species will take food from at least two sources, especially under unfavourable conditions. Identification of the functional group can be made by reference to abdominal colour in live specimens (soil-feeders and soil/wood interface-feeders are darker), site of discovery (in wood, in soil, etc.) and other aspects of biology such as nesting site (arboreal nesters are usually non-soil-feeders), absence of soldiers (generally indicates soil-feeders) and taxonomic affiliation (e.g. all Macrotermitinae are non-soil-feeders; most Apicotermitinae are soil-feeders or soil/wood interface-feeders). If allocation to functional group is difficult, a useful approximation is to divide the species into 'soil-feeders' (soil-feeders and soil/wood interface-feeders as defined above) and 'non-soil-feeders' (all the rest). The distribution of species between these categories indicates termite community structure; forest communities are often dominated by soil-feeders, but disturbance or drying-out generally increases the proportions of other functional groups.

Similarly, it is possible to compare termite assemblages on the basis of nesting types by allocating species to the following categories:

- *Wood nesting*. Termites whose colony centres are normally within dead logs or standing trees. Sometimes the dead wood is gradually replaced with carton material or fungus comb.
- *Hypogeal nesting*. Termites whose colony centres are below the ground. The centres are often poorly defined and amorphous (especially in the soldierless Apicotermitinae), with little obvious internal structure, although some have complex underground nests (e.g. many Macrotermitinae). This group includes many species that are facultative secondary inhabitants of epigeal mounds.
- *Epigeal mounds*. Termites whose colony centres are above ground (but excluding arboreal mounds), free-standing or associated with tree buttresses. These mounds are usually well defined and highly complex structures, built to species-typical specifications but with a tendency to become more irregular as they age through erosion, additions and occupation by secondary inhabitants.
- *Arboreal mounds*. Nests attached to trees at various heights, usually made of carton. In most cases the nests are connected to the ground by covered runways, which may assist in distinguishing some arboreal termite nests from those of ants.

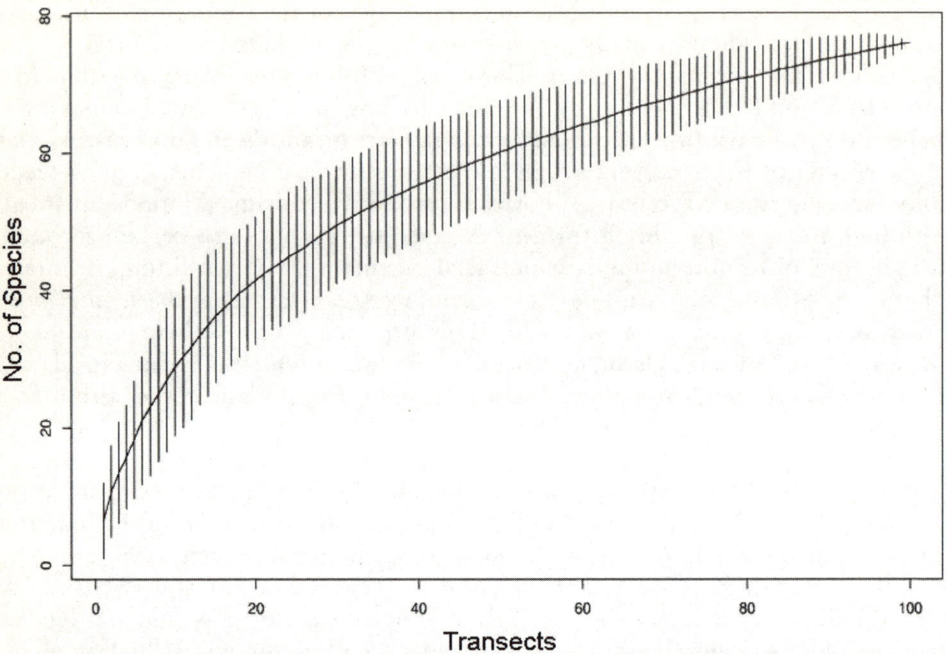

Figure 3.5 *Example of a species accumulation curve with standard deviation using random permutations of the data*

Species accumulation curves

Species accumulation curves are graphs that represent the relation between sampling effort and discovery of species in a habitat or region. The sampling effort may be measured by specimen, sample, trap, time (or person-hours), or area. With a large collecting effort, the curve is asymptotic and tends to reach a plateau. The simplest form of such curves is called the 'collector's curve'. It is just a plot of the accumulated species number against the number of sampling units, in the sequence they were recorded in the field. More sophisticated curves with standard deviations are obtained using random permutations (reiterations, see above) of the data (Figure 3.5) or other methods, which usually require specialized computer programs.

Species accumulation curves can be obtained reiteratively by randomly resampling the soil monoliths or transect sections 500 times using EstimateS software version 6.0b1 (Colwell, 2000). In further statistical analysis, comparison of the mean of the Shannon–Wiener index, species number and (for earthworms) cast production per land use type are performed using the Fisher test at 5 per cent probability level. The Kruskal–Wallis ANOVA can be used to test whether there is a significant effect of land use types on earthworm abundance. The relationship between earthworm abundance and land use index is tested using regressions. Another approach is to normalize earthworm density and biomass before performing the parametric analysis of variance (ANOVA) test. Multivariate analyses like principal component analysis (PCA), correspondence analysis (CA) and co-inertia analyses can be used to determine factors affecting earthworm density and diversity across agricultural landscapes (Thioulouse et al, 1997).

C. Ants and other macrofauna and mesofauna in litter by the Winkler method

Introduction

Ants can be classified in functional groups or guilds. Although some authors consider these names synonymous, they can have different meanings. Guilds are groups of species that utilize the same type of resource for their survival (Terborgh and Robinson, 1986), while functional groups are species that use similar strategies for resource exploitation, and may comprise more than one guild. Several different schemes of functional group classification have been proposed, but we recommend that of Andersen (1997), originally devised for Australian ants, as the most comprehensive:

- *Dominant species*: highly abundant and aggressive species, which influence the distribution and activity of other ant species. Example: *Iridomyrmex* spp.
- *Associated subordinated*: these are mostly species belonging to the genus *Camponotus*, which can coexist with dominant *Iridomyrmex* spp. because of their relatively large body size and submissive behaviour.
- *Climate and soil specialists*: these species (*Melophorus* spp., *Meranoplus* spp., *Aphaenogaster* spp. and *Notoncus* spp.) present physiological, morphological and behavioural adaptations that reduce their interaction with dominant *Iridomyrmex* spp.
- *Cryptic species*: ant species that forage and nest predominantly beneath the ground surface (in the soil and leaf litter), therefore minimizing contact with other ants that do not occupy those habitats (e.g. *Brachyponera* spp., *Eurhopalothrix* spp.).
- *Opportunistic species*: barely competitive, non-specialist ants that frequently thrive in disturbed environments (e.g. *Rhytidoponera* spp., *Paratrechina* spp.).
- *Generalist myrmicines*: species belonging to the genera *Monomorium*, *Crematogaster* and *Pheidole*; they are cosmopolitan, non-specialists, and highly competitive.
- *Solitary foragers of relatively large size*: solitary predators that interact only weakly with other ant species (e.g. species within the genera *Myrmecia*, *Bothroponera* and *Odontomachus*).

The following are examples of guilds in Brazilian Atlantic forest (Delabie, Agosti and Nascimento, 2000):

- *Omnivores species*: the species that use several sorts of food, including the remains of dead animals. Example: *Megalomyrmex*, *Pheidole* and *Solenopsis*.
- *Specialist predators*: species that feed on only one kind of prey. Example: *Amblyopone* (termite predators), *Thaumatomyrmex* (myriapod predators) and *Strumigenys* (collembolan predators).
- *Generalist predators*: species that feed on several kinds of prey. Example: *Gnamptogenys* (ants and other insect predators), *Anochetus* and *Hypoponera*.
- *Legionary ants*: also known as 'formigas da correição', are generalist or specialist predator species. Example: *Eciton*, *Labidus* (generalist predators) and *Neivamyrmex* and *Normamyrmex* (specialist predators; they predate other ants).

- *Soil predators*: species that are established in soil and litter, and forage through it: *Pachycondyla* and *Centromyrmex* (inquilines in termite nests), probably predating the larvae of termites of the genus *Syntermes* (Delabie, 1995).
- *Underground ants*: dependent on honeydew and species that feed on the sugar secretion of other insects. Example: *Acropyga*.
- *Dominant arboreal ants*: omnivorous species nesting in trees but foraging throughout the above-ground vegetation and in the litter. Highly aggressive, having a strong competitive influence over the other species. Example: *Azteca* and *Crematogaster*.
- *Soil or litter dominant ants*: ants that forage though litter and the soil or in the above-ground vegetation. They are subdivided in two groups: a) great generalist predators – *Odontomachus* and *Ectatomma*; b) truly omnivorous – *Brachymyrmex*, *Campanotus*, *Paratrechina*, *Solenopsis* (large species) and *Wasmannia*.
- *Fungus cultivators*: species that utilize symbiotic fungi, usually members of the Attini, using fresh leaves, carcasses and the remains of plants for cultivation of the fungi. Example: *Acromyrmex*, *Apterostigma*, *Atta*, *Cyphormyrmex*, *Mycocepurus*, *Myrmicocrypta*, *Sericomyrmex* and *Trachymyrmex*,

Other functional classifications are possible. For practical fieldwork, short classifications are more convenient, for example that agreed by the CSM-BGBD Workshop (2005):

- carnivores (specialists or generalists) with a subterranean colony centre;
- carnivores (specialists or generalists) with colony centres in the superficial litter layers;
- carnivores (specialists or generalists) with an epigeic or arboreal colony centre;
- herbivores with a subterranean colony centre;
- herbivores with colony centres in the superficial litter layers;
- herbivores with an epigeic or arboreal colony centre.

Sampling procedure

The full process takes place in three stages: a) a crude sieving in the field which separates finer litter (with ants) from coarse material; b) transport to the laboratory in a secure bag; c) extraction of the animals inside the closed but aerated Winkler system. Litter is first collected from each of at least three $1m^2$ quadrats, preferably arranged along a transect line adjacent to the sampling GPS coordinate and the monolith (Plate 1, see also Chapter 2). A square frame is used to delimit the $1m^2$ area and litter is removed with the hands, working from the rim to the centre, wearing leather gloves. Each sample of litter is sieved to remove coarse items by shaking the sieve sideward and upwards to separate the thicker fragments. The finer fraction is collected at the bottom of the sieve bag, together with the ants, which is then transferred to the field bags. This is done expeditiously to avoid insect escape. In the laboratory the contents of the field bags are transferred into the net bags that are subsequently fixed inside the Winkler extractors that are then hung under ambient conditions under cover in a well-aired place. A cup containing 70 per cent alcohol is fitted to the base of the Winkler extractor. As the litter slowly dries out, ants and other insects migrate outwards and fall through to the alcohol. Extraction is maintained for at least 72 hours, to maximize recovery of specimens (Delabie, 1999). Specimens are stored in closed labelled tubes with 70 per cent alcohol. As an addition, or an alternative if Winklers are not available, coarsely sieved litter can be brushed through a 2mm mesh onto a white cloth sheet. This separates smaller macrofauna and some mesofauna from the organic material, and the animals can then be collected from the sheet using a hand-held aspirator (pooter). In this case a search limit of 30 minutes should be observed per $1m^2$ of litter sampled. Dried litter from the Winkler bags should be weighed at the conclusion of the extraction.

The full equipment list for the Winkler extractor (Bestelmeyer et al, 2000) is as follows:

- Square frame, made of wood, plastic or aluminium, measuring 1.0×1.0m, used to delimit the sample area.
- Coarse sieve, of about 5mm, fitted medially inside a bag made of a resistant synthetic nylon or tergal, approximately 1.5m long.
- Field bags, of similar material measuring approximately 34cm width and 55cm length, which are used to transport crudely sieved material from the field to the locations where the extractors are placed.
- Net bags, one per extractor, 25cm wide and 37cm long which are loosely filled with the litter transferred from field bags, and positioned centrally within the extractor such that the invertebrates can migrate through.
- Winkler extractor itself, consisting in an insect-proof but porous cotton or synthetic nylon bag, measuring 1.50m in length, 25cm width, and 20cm (central) depth. The bottom of the bag tapers to a plastic container of approximately 200mL capacity, containing 70 per cent alcohol.

D. Beetles and other macrofauna and mesofauna in litter: Baited and non-baited pitfalls

Introduction

Pitfall trapping is an effective means of collecting samples of the more mobile litter-dwelling meso- and macro-invertebrates. Pitfall traps are deep relatively small containers (e.g. improvised from jars, plastic bottles, tins, cups), which are buried so that the top of the container is exactly at the level of the ground surface after loose litter has been cleared away at the point where the trap is to be located. In a variant of the placement, the top of the container can be set about 1cm above the ground surface and a ramp of loose soil built up to it on all sides. The traps are covered to exclude rain, and function by collecting ground-dwelling arthropods that fall into the container and are drowned in water at the bottom of the trap. Traps are usually cleared and reset once daily. Pitfalls are cheap and easy to install and widely employed in biodiversity inventory work. Additional advantages are that they sample macrofauna and some mesofauna (especially springtails) simultaneously, and that they can specifically address animals that are active at night. However, there are also disadvantages – principally that they do not sample all taxonomic groups with equal efficiency (beetles, ants, juvenile orthopterans, myriapods, spiders and other large arachnids tend to dominate catches, while winged insects seem able to escape). Pitfall traps are not quantitative and since they are usually deployed in linear arrays (for convenience during setting and clearing), they suffer from the same disadvantage as transects in that there is a high degree of autocorrelation between adjacent samples. Pitfalls are also prone to damage from mammals and birds, and to vandalism by humans.

A typical pitfall trap is illustrated in Figure 3.6. The dimensions are not critical, but the trap should be wide enough to collect a reasonable number of specimens and deep enough to reduce escapes. About 1–2cm of water in the bottom (with a drop of detergent added) drowns or immobilizes most of the animals and the detergent counteracts the hydrophobic properties of many arthropod cuticles. The most usual addition to pitfall traps is a sloped cover to keep out rain. This can be improvised in the field, but an upturned glass Petri dish is ideal, supported on twigs or nails (galvanized or wooden nails are best), to raise the cover about 2–3cm and allow access to the trap. Glass is heavier than plastic, which makes the cover more stable. The cover also creates a cave-like microhabitat that is attractive to surface-active diurnal arthropods with a negative response to light, and helps to keep the contents cool, thus retarding decomposition. It is argued that in direct sunlight the cover should be opaque to avoid a greenhouse effect inside the trap, and light grey in colour so as not to present an attraction to flying insects which are not part of the soil fauna. The cover may also be made of ceramic tile, plastic, wood or other material heavy enough not to be dislodged by wind, or affected by rain or sun, and unpalatable to termites. In some variations of the procedure, salt is added as preservative, but alternatively chloral hydrate solution (5 per cent solution) or lactic acid/acetic acid mixture (10 per cent each, with some glycerine added) can be used to improve preservation.

Pitfall traps can be used in any ground level substrate except water and rock. They can be used to sample arthropods in animal burrows, litter, rotting wood and peatlands, but the most common application is in a soil substratum for surface-active arthropod fauna (Figure 3.7). The use of brightly coloured or patterned containers may improve the catch of parasitoids or other flying insects attracted to colours, but should perhaps otherwise be avoided.

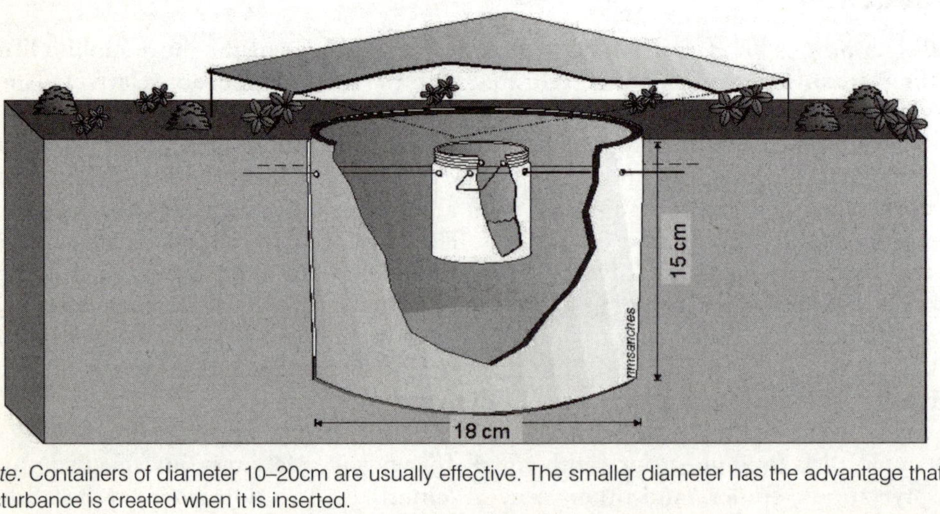

Note: Containers of diameter 10–20cm are usually effective. The smaller diameter has the advantage that less disturbance is created when it is inserted.

Figure 3.6 *A schematic baited pitfall trap*

Note: There is about 2cm of collecting fluid in the bottom of the cup. Top left shows the transect line, marked with a white cord.

Figure 3.7 *Improvised pitfall trap made from a beverage cup, four small sticks and a cellophane film*

Placement of traps

While pitfalls are strictly non-quantitative, they can be useful for the estimation of alpha diversity. In species-richness studies the number of species obtained by pitfall traps will be increased if traps are placed in as many phyto-physiognomies as possible. In studies that are focused on abundance, an approximate surrogate (relative trappability) can be obtained by replicating pitfall traps within a range of phyto-physiognomies. The number of replicates can be based on species accumulation curves and selected to achieve 85 per cent or more of the target species. In other words, what would be the appropriate number of pitfall traps necessary to collect 85 per cent or more of the trappable species in a given phyto-physiognomy? In transect-based sampling (Swift and Bignell, 2001), pitfalls are normally placed in a parallel line on either side, at a distance of about 5m. The number is not specified, and depends on the length of the transect. In the sampling regime recommended for CSM-BGBD (Chapter 2) it is suggested that pitfalls can be placed along the transect line and to the side of the 1m^2 quadrates for litter sampling. The number recommended is three to five per sampling point, but this is not intended to be prescriptive. When a plot is botanically complex and more than one phyto-physiognomy can be recognized, pitfalls can be distributed in the same manner as casual sampling for ants and termites – that is, any microhabitat of interest can be investigated. A variant scheme of pitfall sampling suggested by Southwood (1978) makes use of wooden boards or some similar obstructions up to 30cm high, placed at the surface of the ground in a V or funnel shape, channelling mobile animals to the base of the V or funnel, where an array of closely packed pitfalls can be located (see 'drifting fences', below).

The yield from pitfall trapping is sensitive to disturbance, especially trampling. A sensible precaution is therefore to sequence the entire site – sampling in the order of decreasing vulnerability: pitfall traps first (perhaps overnight), followed in order by transects, mesofaunal cores, microsymbiont cores and finally the monolith. Casual sampling away from the main area of activity can be done at any time.

Installation of pitfall traps

Pitfall traps should be sunk into the ground with the top rim of the container at the level of the soil or litter surface.

1. Use a trowel to cut through and remove the root mat and soil to create a hole which should closely approximate the size of the plastic cup. Care should be taken in removing soil and root material so as to minimize disturbance in the adjacent microhabitat.
2. The pitfall trap is placed in the hole and the outside surrounding edge filled with soil from the excavation and carefully worked flush with the rim. The fill can be used to provide a surface (ramp) for animals to walk unimpeded up to the rim of the trap.
3. Fine soil that falls into the trap during placement can be blown out; larger particles must be removed by hand.
4. Insert three cover supports into the soil surrounding the trap leaving 2.5cm of each support visible above the substratum to support the cover.
5. Fill the traps about one-quarter with water and enough salt to make a saturated solution (if saline is chosen). A few drops of detergent are added as a surfactant. Salt or another preservative is always needed in tropical climates unless the servicing frequency is 24 hours or less. Detergent facilitates the immobilization (by drowning) of active animals.
6. Place the cover on the supports.
7. Place a marker containing the site and trap numbers (in waterproof ink or pencil) near the trap. A stick placed vertically in the ground with a yellow or orange flag of tape attached to the top is useful to identify the location of the trap and to prevent accidental trampling.

Trap servicing and specimen processing

Service traps by removing the cover and lifting the inner cup containing the bait (if present, see below), from the outer cup. Remove large objects such as leaves, twigs, larger animals and so on that may have fallen into the trap. The remaining contents of the cup are gently swirled to lift specimens off the bottom then poured through an aquarium net. To retain collembolans, a separate 53μm sieve is employed.

Specimens are transferred to plastic bags from the net using a wash bottle of water to remove the last specimens. Each trap should be processed separately and a label, containing the site and trap identifiers and date of collection, placed in the bag. Specimens should be processed on the same day in a lab or base camp by rinsing the contents of each bag in the net under a gentle stream of water for several minutes to remove dirt, salt and detergent. Pick out any remaining large debris and attempt to remove as much soil and other organic debris as possible.

Washing of specimens in fresh water is essential because the detergent and dirt will form a film on the insects if they are placed directly in alcohol. This film is difficult or impossible to remove once specimens are preserved, thus rendering them more difficult to image and manipulate under the microscope. After washing, invert the net contents into the specimen bag and, using a squeeze bottle filled with alcohol, gently wash the contents out of the net and down the sides of the specimen bag to the bottom. Cover the sample in the specimen bag with alcohol to at least the level of the sample. Sample labels (written in pencil or waterproof ink) must be placed inside the bag with the specimens. Change the alcohol in about seven days, and if possible, refrigerate or freeze the samples to awaiting further processing. Samples in storage should be double bagged; 95 per cent alcohol is used for collembolans, 70 per cent alcohol for other specimens.

In variants of the procedure, two collembolan filters (53µm and 38µm mesh) can be used sequentially to separate these animals from other fauna (see Chapter 4). Specimens should be washed to remove detergent and debris, and then preserved in 70 per cent ethanol. For processing of collembolan specimens, see Chapter 4 (Mesofauna). Additional (optional) pitfall traps can be prepared, containing 120mL of 95 per cent ethanol + glycerine (9:1), and left for 72 hours. Two such traps, plus three of conventional design, are recommended per sampling point.

Improving pitfall trap capture capacity

Pitfall trap capture capacity can be improved by several alterations on the trap design or use of baits attractive to arthropods. The main modification on the pitfall design is the enlargement of its diameter, because the pitfall capture capacity is a function of its circumference. The use of attractant baits is a common improvement to pitfall traps to capture a wide range of arthropods. The bait can be dung, carcass, fallen fruit or decaying fungi, which are variously attractive to detritivores and some other types of arthropod, coloured bait (yellow, red or blue) to attract flies or parasitoids, or synthetic attractive substances (e.g. pheromones). Fresh human dung is attractive to a range of beetles and is, of course, easily obtainable. Baits must be separated from the preservative fluid: this can be achieved by suspending the bait in a small pot fitted across the mouth of the main container. The use of baits is controversial in that it biases the sample towards animals with marked olfactory behaviours. Baited traps cannot be used for semi-quantitative assessment such as trappability (for example the percentage of all traps occupied by a given species), because the area from which the sampling has been carried out cannot be determined and the catch is a function not only of animal abundance but also relative motility. However, they can add to locality species lists.

List of materials required

- pitfall traps: 1000mL plastic cups, dimensions: 180mm top diameter ×120mm depth;
- cover: 200×200mm (1 per trap), tile colour: light gray;
- cover supports: 152mm (6-inch) galvanized or wooden nails;
- water: tap water or water from a natural source;
- water bucket or container to carry water;
- pure detergent: unscented liquid dishwashing detergent;
- salt;
- plastic or wooden markers;
- aquarium net or similar, fine mesh;
- sample bags with twist ties. Self-sealing plastic bags (i.e. whirlpack type) are most convenient for field specimen storage;
- alcohol, 70–90 per cent;
- soft-lead pencil;
- card stock for labels.

E. Recording and expressing the data

Parameters

Record the number and fresh weight of all animals and identify to at least the highest resolution of taxonomic and functional levels possible. The presence and weight of termite fungus combs (if any) should also be noted. If a balance is not available in the field, fresh weight can be approximated for preserved specimens by weighing them after light blotting. Make a list of species, if possible grouped into subfamilies or families. Within each of these, use generic names to generate alphabetical orders. Combine the results from monoliths, transects, Winkler quadrats, pitfall traps and casual sampling to compile this list.

Fully identified species should be listed with the full binomial and descriptive authority:

e.g. *Dorylus laevigatus* Smith

Morphospecies should be listed by letter:

e.g. *Crematogaster* sp. A

Crematogaster sp. B

It is important that letter designations should be consistent between sites, e.g. species A is the same morphospecies across the entire benchmark.

Species identified only to genus should be listed without letters:

e.g. *Colobobsis* sp.

Incorporate the species list into a table showing the sites where each occurred.

Table 3.2 *Termite numerical density in seven sites across a forest disturbance gradient in Jambi province, central Sumatra*

Site	*Arithmetical mean, nos m^{-2} (n=5)*	*Geometric mean, nos m^{-2} (n=5)**	*95% confidence limits**
BS1, Primary forest	2892	971	347-12827
BS3, Logged over	163	65	22-977
BS6, *Paraserianthes*	512	47	5-16445
BS8, Rubber	128	11	2-1046
BS10, Jungle rubber	211	25	2-9772
BS12, Alang-alang	3	2	0-20
BS14, Cassava	26	10	2-534
Stratum level (all sites averaged)	*Arithmetical mean, nos m^{-2} (n=5)*	*Geometric mean, nos m^{-2} (n=5)**	*95% confidence limits**
Litter	46	15	3-64
0–10cm	106	80	43-148
10–20cm	55	44	24-78
20–30cm	49	4	1-50

Note: * back-transformed.
Source: Specimen data after Jones et al, 2003

Estimate abundance as numbers per m^2, from each monolith and each stratum (multiply the raw number per monolith by 16 (except earthworms and millipedes, which are sampled from the monolith and the trench, see 'monoliths' above). Determine abundance for individual taxa, as well as higher taxonomic groups and by combining data for all species. Calculate arithmetical means. To estimate the 95 per cent confidence limits the primary data should be transformed as $\log_{10}(x+1)$. If there are not too many zeros (samples with no animals), this should roughly normalize the data and produce homogeneous variances from group to group. In difficult cases a loglog transformation can be tried. Apply descriptive statistics to the transformed dataset, including 95 per cent confidence limits, and back transform to obtain a geometric mean. Quote means for untransformed data, together with (back-transformed) geometric means and confidence limits for $\log_{10}(x+1)$ transformed data. The transformed data can be used for histograms and site-to-site comparisons (Eggleton et al, 1996). Prepare a summary table, as given in the example in Table 3.2.

Some indication of relative abundance may be obtained from the frequency of encounter of individual species (number of sections of a transect where they were found, out of the total number sampled, number of pitfalls out of the total set, etc.). Also, in the sense that a whole transect represents a single large sample, it is arguably possible to assign dominance and therefore to derive indices of species diversity and equitability. It is also possible that the overall number of encounters – that is, the number of times separate colonies (for ants and termites), or for other groups separate individuals, are discovered in the transect as a whole – may be a useful population parameter. This concept cannot so easily be applied to pitfalls. For the full range of information obtainable from transects, see the discussions in Eggleton and Bignell (1995), Eggleton et al (1997), Jones and Eggleton (2000), Woodcock (2005) and Jones et al (2005).

Estimate biomass as g m^{-2} in a similar way as abundance. Use fresh weight or the mass of blotted preserved specimen, if possible. Avoid the use of dry weight because of the different oven temperatures used by different scientists and the variable water content of different types of organism. Where insect specimens in a range of sizes are available, an alternative method is to calibrate live biomass against head width in representative specimens covering the whole size range. The weight of unknowns can then be estimated from the curve. For log transformations of data, it is most convenient to work in (mg + 1), and then back-transform and express as g. Prepare a summary table, as above.

NB: the determination of abundance only is easier, as it does not involve the weighing of individual specimens, or other measurements to estimate biomass; however, many ecosystem process rates are generally proportional to biomass, not abundance. Hence, survey data without biomass estimates are incomplete and less informative. Show species/morphospecies richness, abundance and biomass graphically.

Analysis

Analyses should be conducted per functional group, that is, defined by feeding habit and distribution in the soil profile. For comparability, the classification for earthworms (already given) can be applied to all soil invertebrates:

Epigeic species, which live and feed on the soil surface. These invertebrates effect litter comminution and nutrient release, but do not actively redistribute plant materials (though the comminuted material may be more easily transported by wind or water than the material from which it was derived). This group includes mainly a variety of arthropods (viz. ants, beetles, cockroaches, centipedes, millipedes, woodlice and orthopterans), gastropods and small, entirely pigmented earthworms. Surface-active macrofauna will include those organisms sampled by pitfall traps.

Anecic species, which remove litter from the soil surface through their feeding activities. Considerable amounts of soil, mineral elements and organic matter may be redistributed through these activities, accompanied by physical effects on soil structure and hydraulic properties. Earthworms and non-soil-feeding termites are the main groups in this category, but also some arachnids.

Endogeic species, which live in the soil and feed on organic matter and dead roots, also ingesting large quantities of mineral material. The two main groups are earthworms and soil-feeding termites.

Assemblages can be compared by the relative proportions of species or recognizable taxonomic units (morphospecies), which can be allocated to one (or more) of the functional categories. It is recommended that an additional functional group classification can be attempted for macrofauna, using the following categories:

Ecosystem engineers (usually ants, termites, anecic and endogeic earthworms). Large invertebrates (>1cm, but sometimes smaller) ingesting a mixture of organic material and mineral debris, forming stable, long-lived faeces which are organo-mineral complexes.

Keystone species (e.g. termites and arguably some litter transformers) providing physical niche opportunities for lower-level organisms and determining the community structure of those organisms. NB: the terms 'ecosystem engineer' and 'keystone species' are sometimes used synonymously, but this is incorrect.

Litter transformers (usually non-social or perhaps semi-social arthropods, epigeic earthworms and the larger mites) ingesting a mixture of organic matter and microbial biomass and forming short-lived holorganic faeces. This group includes large-sized (>1cm) and medium-sized (0.2–1cm) invertebrates, but some smaller animals (e.g. many mites) may have analogous functions.

Macropredators (predatory species >0.2cm, usually from various arthropod groups). Actual species richness may be estimated by Sob, ACE, ICE and EstimateS. Useful diversity and similarity indices include Shannon–Wiener, Simpson, Jackknife, Jacquard and Sorensen (see Southwood, 1978 and Krebs, 1998).

F. Minimum data sets

Minimum datasets can be expressed per point sample, per land use category and per sampling window as given below.

Minimum data set per point sampled

From all sampling

Species/morphospecies lists for ants, termites, earthworms and baited beetles.
Family list for other beetles.
List of other invertebrates, to highest taxonomic resolution possible.
For each taxon, please indicate the sampling method employed as:

C casual
T transect
W Winkler
P pitfalls
M monolith

(for the monolith, don't observe 3×10cm stratification, except for earthworms).

Total abundance as nos m^{-2}, separately for:

- all ants (plus per functional group (FG1, FG2, etc.);
- all termites (plus per FG1, FG2, etc. or by default as non-soil-feeders and soil-feeders);
- all earthworms (plus per FG1, FG2, etc.);
- all beetles;
- other invertebrates;
- all invertebrates.

From the transect or transects

Relative abundance of termites. The relative abundance for each FG is given by the number of encounters divided by the total number of encounters for all FGs.

Make one calculation for each FG, using all transect sections available for the sampling. Note that one encounter is occurrence in one 5m section of the transect. The number of encounters therefore varies from zero to four for 20m transects and from zero to 20 for a 100m transect. Data from different transects run at separate sampling points can be pooled.

From the Winkler bags

Frequency of species per sample.
Abundance data for ants and beetles (nos m^{-2}). For ants total numbers per functional group (as FG1, FG2, etc.) can also be estimated.

From the pitfall traps

Itemize taxa not sampled by other means (only).

Minimum data sets per land use

The land use categories are scored or, when this is not possible, ranked according to land use intensity.

From all sampling

Species/morphospecies lists for ants, termites, earthworms and baited beetles.
Family list for other beetles.
List of other invertebrates, to highest taxonomic resolution possible.
For each taxon, please indicate the sampling method employed using the same codes as specified earlier: 'C' for casual, 'T' for transect, 'W' for Winkler, 'P' for pitfalls and 'M' for monolith.

From the monoliths

Provide mean abundance as:
a) Nos m^{-2} ± SD
b) Geometric mean ± 95 per cent confidence interval and
c) Arithmetic mean ± 95 per cent confidence interval, where n = no of points per land use, separately for all ants (and for functional group FG1, FG2, etc.), all termites (and for FG1, FG2, etc. or by default as wood-feeders and soil-feeders), all earthworms (and per FG1, FG2, etc.), all beetles, other invertebrates and all invertebrates (the same as indicated for minimum data per point sampled).

From the transects

Mean relative abundance of termite FG1, FG2, etc. as percentage ± SD, where n = no of points per land use.

From the Winkler bags

Mean abundance of ants (and as FG1, FG2, etc.) as nos m^{-2} ± SD, where n = no of points per land use.
Mean abundance of beetles as nos m^{-2} ± SD.

Minimum data sets per sampling window

From all sampling methods

Species turnover along the gradient as Whittaker's β, i.e. β-diversity.

For all mean total abundances and mean FG relative abundances

Correlates with individual gradient parameters, including land use intensification scores.

The use of raw abundance and biomass data in the interpretation of ecosystem function is controversial. For example, it is argued that abundance of individuals should not be used for ants because they are social insects that organize themselves in colonies with thousands of individuals and which feed in groups. If the sampling is made on a colony the back-calculation to numerical density will be overestimated (Colwell and Coddington, 1994). The frequency of species by sample is an alternative parameter. Similarly, it is argued that organisms should only be designated as ecosystem engineers when their abundance or biomass justifies this, on a site by site basis. The counter-arguments are that social insects can be accurately assessed for their abundance and biomass if the sampling regime is sufficiently robust (including adequate replication, nested sampling and stratification, e.g. Eggleton et al, 1996) and that biomass is of such overwhelming importance in determining ecological impact that it cannot be overlooked or substituted. Unfortunately, the scale of effort required to assess biomass is large (see Eggleton and Bignell, 1995; Lawton et al, 1998) and this compromises attempts to characterize landscape heterogeneity by restricting the number of sampling points that can be used.

References

Agosti, D., Majer, J. D., Alonso L. T. and Schultz T. (eds) (2000) *Measuring and Monitoring Biological Diversity. Standard Methods for Ground-living Ants*, Smithsonian Institution, Washington, DC

Andersen, A. N. (1997) 'Functional groups and patterns of organization in North American ant communities: A comparison with Australia', *Journal of Biogeography*, vol 24, pp433–460

Anderson, J. M. and Ingram, J. S. I. (eds) (1989) *Tropical Soil Biology and Fertility: A Handbook of Methods*, 1st edition, CAB International, Wallingford

Anderson, J. M. and Ingram, J. S. I. (eds) (1993) *Tropical Soil Biology and Fertility: A Handbook of Methods*, 2nd edition, CAB International, Wallingford

Barros, E., Pashanasi, B., Constantino, R. and Lavelle, P. (2002) 'Effects of land-use system on the soil macrofauna in western Brazilian Amazonia', *Biology and Fertility of Soils*, vol 35, pp338–347

Bestelmeyer, B. T., Agosti, D., Leeanne, E., Alonso, T., Brandão, C. R. F., Brown, W. L., Delabie, J. H. C., Bhattacharya, T., Halder, G. and Saha, R. K. (2000) 'Soil microarthropods of a rubber plantation and a natural forest', *Environmental & Ecology*, vol 3, pp143–147

Bhadauria T., Ramakrishnan, P. S. and Srivastava, K. N. (2000) 'Diversity and distribution of endemic and exotic earthworms in natural and regenerating ecosystems in the central Himalayas, India', *Soil Biology and Biochemistry*, vol 32, pp2045–2054

Blakemore, R. J. (2002) *Cosmopolitan Earthworms – An Eco-Taxonomic Guide to the Peregrine Species of the World*, VermEcology, Kippax, Australia

Blakemore, R. J. (2005) 'Introductory key to revised earthworm families of the world', in N. Kaneko and M.T. Ito (eds) *A Series of Searchable Texts on Earthworm Biodiversity, Ecology and Systematics from Various Regions of the World*, COE Soil Ecology Research Group, Yokohama National University, Japan, www.bio-eco.eis.ynu.ac.jp/eng/index.htm

Blanchart, E., Lavelle, P., Braudeau, E., Lebissonnais, Y. and Valentin, C. (1997) 'Regulation of soil structure by geophagous earthworm activities in humid savanna of Côte d'Ivoire', *Soil Biology and Biochemistry*, vol 29, pp431–439

Bouché, M. M. (1977) 'Stratégies lombriciennes', in U. Lohm and T. Persson (eds) *Soil Organisms as Components of Ecosystems*, *Ecological Bulletins*, vol 25, pp122–132

Burel, F., Baudry, J., Butet, A., Clergeau, P., Delettre, Y., Le Coeur, D., Dubs, F., Morvan, N., Paillat, G., Petit, S., Thenail, C., Brunel, E. and Lefeuvre, J.-C. (1998) 'Comparative biodiversity along a gradient of agricultural landscapes', *Acta Oecologia*, vol 19, pp47–60

Collins, N. M. (1989) 'Termites', in H. Leith and M. J. A. Werger (eds) *Tropical Rain Forest Ecosystems*, Elsevier Science Publishers, Amsterdam

Colwell, R. K. (2000) 'Statistical Estimate of Species Richness and Shared Species from Samples, Version 6.0b1', http://viceroy.eeb.ucon.edu/estimates

Colwell. R. K. (2005) 'EstimateS, Version 7.5: Statistical Estimation of Species Richness and Shared Species from Samples. Software and User's Guide', http://viceroy.eeb.uconn.edu/EstimateS

Colwell, R. K. and Coddington, J. A. (1994) 'Estimating terrestrial biodiversity through extrapolation', *Philosophical Transactions of the Royal Society (Series B)*, vol 345, pp101–118

CSM-BGBD (2005) *Global workshop on ant and termite sampling and identification*, Tropical Soil Biology and Fertility Institute of CIAT, Nairobi, Kenya

Csuzdi, C. S. (1996) 'Revision der Unterfamilie Benhamiinae MICHAELSEN, 1897 (Oligochaeta: Acanthodrilidae)', *Mitteilungen aus dem Zoologischen Museum in Berlin*, vol 72, pp347–367

Decaëns, T. and Jiménez, J. J. (2002) 'Earthworm communities under an agricultural intensification gradient in Colombia', *Plant and Soil*, vol 240, pp133-143

Decaëns, T., Jiménez, J. J., Barros, E., Chauvel, A., Blanchart, E., Fragoso, C. and Lavelle, P. (2004),'Soil macrofauna communities in permanent pastures derived from tropical forest or savannas', *Agriculture, Ecosystems and Environment*, vol 103, pp301–312

Decaëns, T., Lavelle, P., Jiménez, J. J., Escobar, G. and Rippstein, G. (1994) 'Impact of land management on soil macrofauna in the Oriental Llanos of Colombia', *European Journal of Soil Biology*, vol 30, pp157–168

Delabie, J. H. C. (1995) 'Inquilismo simultâneo de duas eespécies de *Centromyrmex* (Hymenoptera. Formicidae, Ponerinae) em cupinzeiros de *Syntermes* sp. (Isoptera, Termitidae, Nasutermitinae)', *Revista Brasileira De Entomologia, São Paulo*, vol 39, pp605–609

Delabie, J. H. C. (1999) 'Comunidades de formigas (Hymenoptera: Formicidae): Métodos de estudo e estudos de casos na Mata Atlântica', *Encontro De Zoologia Do Nordeste*, vol 12, *Feira De Santana*, Resumos Feira De Santana, Sociedade Nordestina De Zoologia, pp58–68

Delabie, J. H. C., Agosti, D. and Nascimento, I. C. (2000) 'Litter ant communities of the Brazilian Atlantic rain forest region', in D. Agosti, J. D. Majer, L. T. Alonso and T. Schultz (eds) *Measuring and Monitoring Biological Diversity: Standard Methods For Ground-living Ants*, Smithsonian Institution, Washington, DC

DeSouza, O. F. F. and Brown, V. K. (1994) 'Effects of habitat fragmentation on Amazonian termite communities', *Journal of Tropical Ecology*, vol 10, pp197–206

Dlamini, C., Haynes, R. J. (2004) 'Influence of agricultural land use on the size and composition of earthworm communities in northern KwaZulu-Natal, South Africa', *Applied Soil Ecology*, vol 27, pp77–88

Easton, E. G. (1979) 'A revision of the 'acaecate' earthworms of the *Pheretima* group (Megascolecidae: Oligochaeta): *Archipheretima*, *Metapheretima*, *Planapheretima*, *Pleinogaster* and *Polypheretima*', *Bulletin of the British Museum (Natural History) Zoology*, vol 35, pp1–128

Eggleton, P. and Bignell, D. E. (1995) 'Monitoring the response of tropical insects to changes in the environment: troubles with termites', in R. Harrington and N. E. Stork (eds) *Insects in a Changing Environment*, Academic Press, London

Eggleton, P. and Bignell, D. E. (1997) 'The incidence of secondary occupation of epigeal termite (Isoptera) mounds by other termites in the Mbalmayo Forest Reserve, Southern Cameroon, and its biological significance', *Journal of African Zoology*, vol 111, pp489–498

Eggleton, P., Bignell, D. E., Sands, W. A., Mawdsley, N. A., Lawton, J. H., Wood, T. G. and Bignell, N. C. (1996) 'The diversity, abundance and biomass of termites under differing levels of disturbance in the Mbalmayo Forest Reserve, Southern Cameroon', *Philosophical Transactions of the Royal Society of London, Series B*, vol 351, pp51–68

Eggleton, P., Homathevi, R., Jeeva, D., Jones, D. T., Davies, R. G. and Maryati, M. (1997) 'The species richness and composition of termites (Isoptera) in primary and regenerating lowland dipterocarp forest in Sabah, East Malaysia', *Ecotropica*, vol 3, pp119–128

Eggleton, P., Homathevi, R., Jones, D. T., MacDonald, J. A., Jeeva, D., Bignell, D. E., Davies, R. G. and Maryati, M. (1999) 'Termite assemblages, forest disturbance and greenhouse gas fluxes in Sabah, East Malaysia', *Philosophical Transactions of the Royal Society of London, Series B*, vol 354, pp1791–1802

Eisen, G. (1900) 'Researches in American Oligochaeta with especial reference to those of the Pacific Coast and adjacent islands', *Proceedings of the Californian Academy of Sciences,* vol 2, pp87–276

Fragoso, C. and Rojas-Fernandes, P. (1994) 'Earthworms from Southeastern Mexico. New Acanthodriline genera and species (Megascolecidae, Oligochaeta)', *Megadrilogica*, vol 6, pp1–12

Fragoso, C., Brown, G. G., Patron, J. C., Blanchart, E., Lavelle, P., Pashanasi, B., Senapati, B. and Kumar, T. (1997) 'Agricultural intensification, soil biodiversity and agroecosystem function in the tropics: The role of earthworms', *Applied Soil Ecology*, vol 6, pp17–35

Fragoso, C., Lavelle, P., Blanchart, E., Senapati, B. K., Jiménez, J. J., Martinez, M. A., Decaëns, T., and Tondoh, J. (1999) 'Earthworm communities of tropical agroecosystems: Origin, structure and influence of management pratices', in P. Lavelle, L. Brussard and P. F. Hendrix (eds) *Earthworm Management in Tropical Agroecosystems*, CAB International, Wallingford

Gates, G. E. (1972) 'Burmese earthworms, an introduction to the systematics and biology of Megadrile oligochaetes with special reference to South-East Asia', *Transactions of the American Philosophical Society*, vol 62, pp1–326

Gay, F. J. and Calaby, J. E. (1970) 'Termites of the Australian region', in K. Krishna and F. M. Weesner (eds) *Biology of Termites*, Academic Press, New York

Gilot, C., Lavelle, P., Blanchart, E., Keli, J., Kouassi, P. and Guillaume, G. (1995) 'Biological activity of soil under rubber plantation in Côte d'Ivoire', *Acta Zoologica Fenica*, vol 196, pp186–189

Greuter, W., McNeill, J., Barrie, F. R., Burdet, H.-M., Demoulin, V., Filgueiras, T. S., Nicolson, D. H., Silva, P. C., Skog, J. E., Trehane, P., Turland N. J. and Hawksworth, D. L. (2000) *International Code of Botanical Nomenclature (St Louis Code)*, Koeltz Scientific Books, Konigstein

Haynes, R. J., Dominy, C. S. and Graham, M. H. (2003) 'Effect of agricultural land use on soil organic matter status and the composition of earthworm communities in KwaZulu-Natal, South Africa', *Agriculture, Ecosystems and Environment*, vol 95, pp453–464

International Commission on Zoological Nomenclature (1999) *International Code of Zoological Nomenclature*, 4th edition, International Trust for Zoological Nomenclature, London

Jones, D. T. and Eggleton, P. (2000) 'Sampling termite assemblages in tropical forests: Testing a rapid biodiversity assessment protocol', *Journal of Applied Ecology*, vol 37, pp191–203

Jones, D. T., Susilo, F.-X., Bignell, D. E., Suryo, H., Gillison, A. W. and Eggleton, P. (2003) 'Termite assemblage collapses along a land-use intensification gradient in lowland central Sumatra, Indonesia', *Journal of Applied Ecology*, vol 40, pp380–391

Jones, D. T., Verkerk, R. H. J. and Eggleton, P. (2005) 'Methods for sampling termites', in S. Leather (ed) *Insect Sampling in Forest Ecosystems*, Blackwell, Oxford

Julka, J. M. (1988) *The Fauna of India and the Adjacent Countries. Megadrile Oligochaeta (Earthworms) Family Octochaetidae*, Zoological Survey of India, Calcutta

Krebs, C. J. (1998) *Ecological Methodology*, 2nd edition, Pearson Education, London

Lavelle, P. (1978) 'Les vers de terres de la savane de Lamto (Côte d'Ivoire): Peuplements, populations et fonction dans l'écosystème', Thèse de Doctorat, Université Paris VI, Laboratoire de Zoologie, Ecole Normale Supérieure 12, France

Lavelle, P. (1981) 'Strategies de reproduction chez les vers de terre', *Acta Oecologia – Oecologia Generalis*, vol 2, pp117–133

Lavelle, P. (1983) 'The soil fauna of tropical savannas. II The earthworms', in F. Bourlière (ed) *Tropical Savannas*, Elsevier Publishing Company, Amsterdam

Lavelle, P., Bignell, D., Lepage, M., Wolters, V., Roger, P., Heal, O. and Dhillion, S. (1997) 'Soil function in a changing world: The role of the invertebrate ecosystem engineers', *European Journal of Soil Biology*, vol 33, pp159–193

Lawton, J. H., Bignell, D. E., Boulton, B., Bloemers, G. F., Eggleton, P., Hammond, P. M., Hodda, M., Holt, R. D., Larsen, T. B., Mawdsley, N. A., Stork, N. E., Srivastava, D. S. and Watt, A. D. (1998) 'Biodiversity inventories, indicator taxa and effects of habitat modification in tropical forest', *Nature (Lond)*, vol 391, pp72–76

Lee, K. E. (1987) 'Peregrine species of earthworms', in A. M. Bonvicini and P. Omodeo (eds) *On Earthworms: Selected Symposia and Monographs*, Unione Zoologica Italiana 2, Mucchi, Modena

Magurran, A. E. (2004) *Measuring Biological Diversity*, Blackwell Science, Oxford

Mathieu, J., Rossi, J.-P., Grimaldi, M., Mora, P., Lavelle, P. and Rouland-Lefèvre, C. (2004) 'A multi-scale study of soil macrofauna biodiversity in Amazonia pastures', *Biology and Fertility of Soils*, vol 40, pp300–305

Moreno, E. C. and Halffter, G. (2000) 'Assessing the completeness of bat biodiversity inventories using species accumulation curves', *Journal of Applied Ecology*, vol 37, pp149–158

Moreno, E. C. and Halffter, G. (2001) 'Spatial and temporal analysis of α, β and γ diversities of bats in a fragmented landscape', *Biodiversity and Conservation*, vol 10, pp367–382

Ortiz-Ceballos, I. A. and Fragoso, C. (2004) 'Earthworm populations under tropical maize cultivation: The effect of mulching with velvet bean', *Biology and Fertility of Soils*, vol 39, pp438–445

Pielou, E. C. (1996) 'The measurement of diversity in different types of biological collections', *Journal of Theoretical Biology*, vol 13, pp213–226

Righi, G. (1971) 'Sobre a familia Glossoscolecidae (Oligochaeta) no Brasil', *Arquivos de Zoologia*, vol 20, pp1–95

Righi, G. (1995) 'Colombian earthworms', *Studies on Tropical Andean Ecosystems*, vol 4, pp485–607

Sims, R. W. (1980) 'A classification and the distribution of earthworms, suborder Lumbricina (Haplotaxida: Oligochaeta)', *Bulletin of the British Museum (Natural History) Zoology*, vol 39, pp103–124

Sims, R. W. (1987) 'A review of the Central African earthworm family Eudrilidae (Oligochaeta)', in A. M. Pagliani and P. Omodeo (eds) 'On earthworms, selected symposia and monographs', *Unione Zoologica Italiana* 2, Mucchi, Modena

Sims, R. W. and Easton, E. G. (1972) 'A numerical revision of the earthworm genus *Pheretima* auct. (Megascolecidae: Oligochaeta) with the recognition of new genera and an appendix on the earthworms collected by the Royal Society North Borneo Expedition', *Biological Journal of the Linnean Society*, vol 4, pp169–268

Sleaford, F., Bignell, D. E., and Eggleton, P. (1996) 'A pilot analysis of gut contents in termites from the Mbalmayo Forest Reserve, Cameroon', *Ecological Entomology*, vol 21, pp279–288

Sneath, P. H. A. (1992) *International Code of Nomenclature of Bacteria: Bacteriological Code, 1990 Revision*, ASM Press, Herndon, VA

Southwood, T. R. E. (1978) *Ecological Methods with Particular Reference to the Study of Insect Populations*, 2nd edition, Chapman and Hall, London

Swift, M. and Bignell, D. (eds) (2001) 'Standard methods for the assessment of soil biodiversity and land-use practice', International Centre for Research in Agroforestry, South East Asian Regional Research Programme, ASB Lecture Note 6B, Bogor, Indonesia, www.fao.org/ag/agl/agll/soilbiod/docs/manual-soil%20bioassessment.pdf

Terborgh, R. and Robinson, S. (1986) 'Guilds and their utility in ecology', in J. E. Kikkawa and D. Anderson (eds) *Community Ecology, Pattern and Process*, Blackwell Science, Oxford

Thioulouse, J., Chessel, D., Dolédec, S. and Olivier, J. M. (1997) 'ADE: A multivariate analysis and graphical display software', *Statistics and Computing*, vol 7, pp75–83

Tondoh, E. J. and Lavelle, P. (2005) 'Population dynamics of *Hyperiodrilus africanus* (Oligochaeta, Eudrilidae) in Ivory Coast', *Journal of Tropical Ecology*, vol 21, pp493–500

Vincent, J. P. (1969) 'Recherche sur le peuplement en Oligochète de la savane de Lamto (Côte d'Ivoire)', Thèse de 3e cycle, Université Paris VI, France

Wood, T. G. (1978) 'Food and feeding habits of termites', in M. V. Brian (ed) *Production Ecology of Ants and Termites*, Cambridge University Press, Cambridge

Woodcock, B. A. (2005) 'Pitfall sampling in ecological studies', in S. Leather (ed) *Insect Sampling in Forest Ecosystems*, Blackwell, Oxford

Zicsi, A. (2001) 'Revision der Untergattung *Martiodrilus* (*Maipure* Righi, 1995) (Oligochaeta: Glossoscolecidae), Regenwürmer aus Südamerika', *Opuscula Zoologica Budapest*, vol 33, pp113–131

Zicsi, A. (2005) 'Über den heutigen Stand der Regenwurmfauna (Oligochaeta) Ekuadors mit einem Bestimmungsschlüssel der Glossoscoleciden-Arten', Regenwürmer aus Südamerika 39, *Berichte des Naturwissenschaftlicher-Medizinischer Vereins in Innsbruck*, vol 92, pp95–130

Appendix 1 Termite, beetle and ant identification keys and sampling protocols per region

Region and author	*Title and reference*	*Notes*
SE ASIAN ANTS		
Hashimoto, Y. and Homathevi, R. (2003)	'Manual for Bornean Ant (Formicidae) Identification', prepared for 'Tools for Monitoring Soil Biodiversity in the ASEAN Region', Universiti Malaysia Sabah, 12–26 October 2003, pp95–162	Unpublished key to ant genera in Borneo (useful in SE Asia, including Indonesia). Well illustrated key with generic synopsis.
NEOTROPICAL ANTS		
Fernandez, F. (2003)	'Ontroducción a las Hormigas de al region Neotropical', Instituto de Investigación de Recursos Biológicos Alexander von Humboldt, Bogotá, Colombia, XXVI, pp1–398	Published key to ant genera in Neotropical Region. Includes a review of the biology, ecology, taxonomy and economic importance of the ants.
SE ASIAN TERMITES		
Thapa, R. S. (1981)	'Termites of Sabah', *Sabah Forest Record 12*, pp1–374	The standard work, but perfectly usable.
Collins, N. M. (1984)	'The termites (Isoptera) of Gunung Mulu National Park with a key to the genera known from Sarawak', *Sarawak Museum Journal*, vol 30, pp65–87	Somewhat outdated key to genera of termites in North Borneo, but generally easy to use.
Tho, Y. P. (1992)	'Termites of peninsular Malaysia', in L. G. Kirton (ed) *Malaysian Forest Records*, vol 36, pp1–244. Forest Research Institute of Malaysia, Kepong	Inexpensive and in print. Well illustrated and easy guide to identification by the soldier caste, but does not cover the full diversity of soil-feeding forms in the *Termes* clade.
Gathorne-Hardy, F. (2001)	'A review of the South-east Asian Nasutitermitinae (Isoptera: Termitidae), with descriptions of one new genus and a new species and including a key to genera', *Journal of Natural History*, vol 35, pp1485–1506	Straightforward modern key, useful for Sumatra, Borneo, Java and Sulawesi.

Region and author	*Title and reference*	*Notes*
Homathevi, R. (2003)	'Manual for Bornean Termites (Isoptera)', prepared for 'Tools for Monitoring Soil Biodiversity in the ASEAN Region', Universiti Malaysia Sabah, 12–26 October 2003, pp2–38	Unpublished key to termite genera in Borneo (useful in SE Asia, including Indonesia). Well illustrated key with many references and incorporating recent taxonomic descriptions.
ASIAN BEETLES		
Chung, A. Y. C. (2003)	'Manual for Bornean beetle (Coleoptera) family identification', prepared for 'Tools for Monitoring Soil Biodiversity in the ASEAN Region', Universiti Malaysia Sabah, 12–26 October 2003, pp1–28	Well-illustrated short modern key to family groups with diagnostic synopsis of more than 100 families and almost 50 subfamilies.
Homathevi, R. (2006)	Training course on Arthropod preservation, curation and data management: Coleoptera, Universiti Malaysia Sabah, 18–22 September 2006	Adaptation from Chung (2003).
GENERAL BEETLES		
Booth, R. G., Cox, M. L. and Madge, R. B. (1990)	*IIE Guides to Insects of Importance to Man. 3, Coleoptera*. International Institute of Entomology (an institute of CAB International), The Natural History Museum, London	An extensive explanation on morphology and groupings in beetles.
AFRICAN TERMITES		
Sands, W. A. (1959)	'A revision of the termites of the genus *Amitermes* from the Ethiopian region', *Bulletin of the British Museum (Natural History)*, vol 8, pp129–156	The current standard authority on this genus. Keys to species from both imagos and soldiers.
Webb, G. C. (1961)	*Keys to the Genera of the African Termites*, Ibadan University Press, Ibadan, Nigeria	Well-illustrated key to genera from the soldiers. Somewhat outdated, but easy to use.
Sands, W. A. (1965)	'A revision of the termite subfamily Nasutitermitinae (Isoptera, Termitidae) from the Ethiopian region', *Bulletin of the British Museum (Natural History)*, Supplement 4, pp1–172	The current standard authority on this subfamily in Africa. Keys to genera from both imagos and soldiers.

Region and author	Title and reference	Notes
Bouillon, A. and Mathot, G. (1966)	'Quel est ce termite Africain?', *Zooleo*, vol 1, pp1–115	Keys to family from the imago and soldier. Key to genera from the soldier, and many genera lavishly illustrated at species level. (In French.)
Ruelle, J. E. (1970)	'A revision of the termites of the genus *Macrotermes* from the Ethiopean region (Isoptera: Termitidae)', *Bulletin of the British Museum (Natural History) Entomology*, vol 24, pp365–444	The current standard authority on this genus. Keys to species from imagos, major soldiers and minor soldiers.
Harris, W. V. (1971)	*Termites, Their Recognition and Control*, Longman, London	Short outdated key to main genera (only), in African context, but good for training.
Mitchell, B. L. (1980)	'Report on a survey of the termites of Zimbabwe', *Occasional Papers of the National Museums of Rhodesia, Series B., Natural Science* 6, pp187–323	Comprehensive treatise, which includes key to genera from soldiers.
Sands, W. A. (1998)	*The Identification of Worker Castes of Termite Genera from Soils of Africa and the Middle East*, CAB International, Wallingford	Authoritative work, most useful for cryptic subterranean termites, especially Apicotermitinae and all soil-feeders. Inherently difficult.
Kambhampati, S. and Eggleton, P. (2000)	'Taxonomy and phylogeny of termites', in T. Abe, D. E. Bignell and M. Higashi (eds) *Termites: Evolution, Sociality, Symbioses, Ecology*, Kluwer Academic Publishers, pp1–23	Includes a synopsis of families and subfamilies in annotated key form and identifies diagnostic features using imagos, workers and soldiers. Not fully usable as a key, but helpful for confirmations.
NEOTROPICAL TERMITES		
Constantino, R. (2002)	'An illustrated key to Neotropical termite genera (Insecta: Isoptera) based primarily on soldiers', *Zootaxa*, vol 67, pp1–40	Comprehensive and lavishly illustrated key; includes details of gut coiling and enteric valve armatures needed to identify some Apicotermitinae, plus 32 references.

Region and author	Title and reference	Notes
UNPUBLISHED TERMITE KEYS AND CHECKLISTS		
Wanyoni, K. (1983)	'Checklist of the Species of Termites (Isoptera) Recorded from East Africa', (J. P. E. C. Darlington, ed) International Centre of Insect Physiology and Ecology	No key, but checklist identifies distribution between Kenya, Uganda, Tanzania and Zanzibar.
Sands W. A. and Williams, R. M. C. (undated)	'Taxonomy of Termites', Centre for Overseas Pest Research.	Basal document, forming the core of many private keys held in national museums. Includes keys to families and subfamilies from imagos and soldiers, and detailed account and definitions of morphological characters used in taxonomy. Includes a comprehensive bibliography.
Williams, R. M. C. (undated)	'A Key to the Genera of the Termitinae by the Soldier Caste', Centre for Overseas Pest Research	Detailed and usable key to genera of West and Central Africa. Makes excellent advanced training.
MISCELLANEOUS DOCUMENTS		
Roonwal, M. L. and Chhotani, O. B. (1989)	*The Fauna of India and Adjacent Countries: Isoptera (Termites)*, vol I, Zoological Survey of India, Calcutta	Volumes I and II contain keys to families, subfamilies, genera and species, both for imagos and soldiers.
Chhotani, O. B. (1997)	*The Fauna of India and Adjacent Countries: Isoptera (Termites)*, vol II. (Family Termitidae), Zoological Survey of India, Calcutta	Volumes I and II cover the whole subcontinent of India, including Sri Lanka.
Fisher, B. (2001)	'Equipment specifications', Department of Entomology, Harvard University	Lists equipment needed for comprehensive ant sampling, with 2001 prices in $US. Unpublished.
Fisher, B. (2001)	'Guidelines for data collection for ants', Department of Entomology, Harvard University	Guidelines for site characterization and notebook-keeping in connection with ant sampling. Unpublished.
Ward, P. S. (2003)	'Ant subfamilies', Entomology Department, UC Davis	Illustrated synopsis of subfamilies, and distributions, with a large bibliography.

Region and author	Title and reference	Notes
Jones, D. T. (ed) (2003)	'Tools for the Rapid Assessment of Soil Invertebrate Biodiversity in the ASEAN Region', prepared for 'Tools for Monitoring Soil Biodiversity in the ASEAN Region', Universiti Malaysia Sabah, 12–26 October, 2003, pp1–39	Comprehensive and modern review of the nature of soil biodiversity and rapid assessment methods, covering ants, termites, soil beetles, earthworms, diplopods, isopterans and Araneae. Large up-to-date bibliography. Compulsory reading!
Anon (undated)	'Glossary of morphological characters used in ant taxonomy', Department of Entomology, Harvard University	Comprehensive but unillustrated. Unpublished.

APPENDIX 2 TAXONOMIC WORKS WITH KEYS TO SPECIES OF SOME NEOTROPICAL GENERA. KEYS TO GENERA ARE GIVEN IN KRISHNA (1961; KALOTERMITIDAE ONLY), MILL (1983) AND CONSTANTINO (1999, 2002)

Genus	*Reference*	*Content*
KALOTERMITIDAE		
Cryptotermes	Bacchus (1987)	world revision with keys
	Scheffrahn and Krecek (1999)	revision of the species of the West Indies, with key to soldiers
Eucryptotermes	Constantino (1997)	diagnosis; distribution maps
Procryptotermes	Scheffrahn and Krecek (2001)	revision of New World species with keys to soldiers and alates
RHINOTERMITIDAE		
Heterotermes	Constantino (2001)	key to soldiers of South American species
TERMITIDAE		
Amitermes	Light (1932)	revision and keys (outdated)
Anoplotermes	Snyder (1926)	key to alates (outdated)
Atlantitermes	Constantino and DeSouza (1997)	key to soldiers
Cornitermes	Emerson (1952)	revision and keys
Cyranotermes	Constantino (1990)	key to soldiers
Cyrilliotermes	Fontes (1985)	key to soldiers
Labiotermes	Araujo (1954)	key to soldiers of *Paracornitermes* (outdated)
	Emerson and Banks (1965)	revision of *Labiotermes sensu stricto*, with keys
	Constantino et al (2006)	revision and keys
Nasutitermes	Holmgren (1910)	revision and keys (outdated; in German)
Neocapritermes	Krishna and Araujo (1968)	revision and keys
Parvitermes	Scheffrahn and Krecek (1993)	key to soldiers of West Indian species (outdated)
Procornitermes	Emerson (1952)	revision and keys
	Cancello (1986)	revision and key to soldiers (in Portuguese)
Rhynchotermes	Fontes (1985)	key to soldiers (outdated)
Rotunditermes	Fontes and Bandeira (1979)	key to soldiers
Syntermes	Emerson (1945)	revision and keys (outdated)
	Constantino (1995)	revision and keys

APPENDIX 3 SPECIALIZED LITERATURE ON NEOTROPICAL TERMITES

Araujo, R. L. (1954) 'Notes on the genus *Paracornitermes* Emerson 1949 with description of two new species (Isoptera, Termitidae, Nasutitermitinae)', *Revista Brasileira de Entomologi*, vol 1, pp181–189

Bacchus, S. (1987) 'A taxonomic and biometric study of the genus *Cryptotermes* (Isoptera: Kalotermitidae)', *Tropical Pest Bulletin*, vol 7, pp1–91

Cancello, E. M. (1986) 'Revisão de *Procornitermes* Emerson (Isoptera, Termitidae, Nasutitermitinae)', *Papéis Avulsos de Zoologia (São Paulo)*, vol 36, pp189–236

Constantino, R. (1990) 'Notes on *Cyranotermes* Araujo with description of a new species (Isoptera, Termitidae, Nasutitermitinae)', *Goeldiana Zoologi*, vol 2, pp1–11

Constantino, R. (1995) 'Revision of the Neotropical termite genus *Syntermes* Holmgren (Isoptera: Termitidae)', *University of Kansas Science Bulletin*, vol 55, pp455–518

Constantino, R. (1997) 'Notes on *Eucryptotermes* with a new species from central Amazonia (Isoptera: Kalotermitidae)', *Sociobiology*, vol 30, pp125–131

Constantino, R. (2001) 'Key to the soldiers of South American *Heterotermes* with a new species from Brazil (Isoptera: Rhinotermitidae)', *Insect Systematics and Evolution*, vol 31, pp463–472

Constantino, R. (2002) 'An illustrated key to Neotropical termite genera (Insecta: Isoptera) based primarily on soldiers', *Zootaxa*, paper 67, pp1–40

Constantino, R. and DeSouza, O. F. F. (1997) 'Key to the soldiers of *Atlantitermes* Fontes 1979, with a new species from Brazil (Isoptera: Termitidae: Nasutitermitinae)', *Tropical Zoology*, vol 10, pp205–213

Constantino, R., Acioli, A. N., Schmidt, K., Cuezzo, C., Carvalho, S. H. and Vasconcellos, A. A. (2006) 'A taxonomic revision of the Neotropical termite genera *Labiotermes* Holmgren and *Paracornitermes* Emerson (Isoptera: Termitidae: Nasutitermitinae)', *Zootaxa*, paper 1340, pp1–44

Emerson, A. E. (1945) 'The Neotropical genus *Syntermes* (Isoptera: Termitidae)', *Bulletin of the American Museum of Natural History*, vol 83, pp427–472

Emerson, A. E. (1952) 'The Neotropical genera *Procornitermes* and *Cornitermes* (Isoptera, Termitidae)', *Bulletin of the American Museum of Natural History*, vol 99, pp475–540

Emerson, A. E. and Banks, F. A. (1965) 'The Neotropical genus *Labiotermes* (Holmgren): Its phylogeny, distribution and ecology (Isoptera, Termitidae, Nasutitermitinae)', *American Museum Novitates*, no 2208, pp1–33

Fontes, L. R. (1985) 'New genera and new species of Nasutitermitinae from the Neotropical region (Isoptera, Termitidae)', *Revista Brasileira de Zoologia*, vol 3, pp7–25

Fontes, L. R. and Bandeira, A. G. (1979) 'Redescription and comments on the Neotropical genus *Rotunditermes* (Isoptera, Termitidae, Nasutitermitinae)', *Revista Brasileira de Entomologia*, vol 23, pp107–110

Holmgren, N. (1910) 'Versuch einer Monographie der amerikanische *Eutermes* – Arten', *Jahrbuch der Hamburgischen Wissenschaftlichen Anstalten*, vol 27, pp171–325

Krishna, K. (1961) 'A generic revision and phylogenetic study of the family Kalotermitidae (Isoptera)', *Bulletin of the American Museum of Natural History*, vol 122, pp303–408

Krishna, K. and Araujo, R. L. (1968) 'A revision of the Neotropical genus *Neocapritermes* (Isoptera, Termitidae, Nasutitermitinae)', *Bulletin of the American Museum of Natural History*, vol 138, pp84–138

Light, S. F. (1932) 'Contribution toward a revision of the American species of *Amitermes* Silvestri', *University of California Publications in Entomology*, vol 5, pp335–414

Mill, A. E. (1983) 'Generic keys to the soldier caste of New World Termitidae (Isoptera: Insecta)', *Systematic Entomology*, vol 9, pp179–190

Scheffrahn, R. H. and Krecek, J. (1993) '*Parvitermes subtilis*, a new subterranean termite (Isoptera: Termitidae) from Cuba and the Dominican Republic', *Florida Entomologist*, vol 76, pp603–607

Scheffrahn, R. H. and Krecek, J. (1999) 'Termites of the genus *Cryptotermes* Banks (Isoptera: Kalotermitidae) from the West Indies', *Insecta Mundi*, vol 13, pp111–171

Scheffrahn, R. H. and Krecek, J. (2001) 'New world revision of the termite genus *Procryptotermes* (Isoptera : Kalotermitidae)', *Annals of the Entomological Society of America*, vol 94, pp530–539

Snyder, T. E. (1926) 'Termites collected on the Mulford Biological Exploration to the Amazon Basin 1921–1922', *Proceedings of the US National Museum*, vol 68, pp1–76

Chapter 4

Soil Collembola, Acari and Other Mesofauna – The Berlese Method

Agus Karyanto, Cahyo Rahmadi, Elizabeth Franklin, F.-X. Susilo and Jose Wellington de Morais

Introduction

Mesofauna comprises animals of body sizes ranging from 0.2 to 2.0mm whose abundance cannot be accurately assessed by means of hand sorting from the soil. Enchytraeidae, Acari (mites), Collembola (collembolans), Protura, Pauropoda and some Nematoda are typical members of the mesofauna, but larval forms of macrofaunal species also fall within the size range of mesofauna. The sampling for mesofauna (and other soil biota) is integrated within the overall strategy for the inventory of below-ground biodiversity as described in Chapter 2. At each sampling point, soil cores are located in two concentric circles, at 3m and 6m radius from the monolith, and supplemented by pitfalls located at least 14m from the monolith centre. Mesofauna is collected by using two sampling methods: a) a pooled sample of soil cores and surface litter is extracted either by Berlese–Tullgren or by a modified Berlese, and b) from the fluid contents of pitfall traps. Soil cores are used primarily to collect mesofauna that inhabit the mineral soil beneath the surface litter and the litter itself, whereas pitfall trap methods are applied to collect soil surface dwellers.

The Berlese method, a funnel-type extraction system for separation of small arthropods from soil was developed by Antonio Berlese at the end of the 19th century. In that system, hot water circulates between double brass walls of the funnel (or funnels), slowly drying the sample. In 1918, Tullgren modified the Berlese method by replacing the water jacket with an electric light bulb. The pitfall method (Chapter 3) is good enough to catch collembolans that live at the surface of the soil (edaphics). It is not useful in collecting oribatid mites (Acari: Oribatida), as these animals are mostly sedentary (except for some families such as Galumnidae and Scheloribatidae), in contrast to more opportunistic groups such as collembolans. It is suggested that for all normal sampling, both funnel-based extractions and pitfalls should be routinely employed together.

Oribatid mites (also called Cryptostigmata, and referred to as beetle mites or box mites) are one of the most numerically dominant (densities can reach several hundred thousand individuals per square metre) and species diverse (undisturbed soil may yield 50–100 species) arthropod groups in the organic horizons of most soils (Norton, 1990). They feed on living and dead plant material and carrion, graze on fungi and algae, and some are predaceous (e.g. feed on live nematodes) (Siepel, 1990). Their faecal pellets provide a large surface area for primary decomposition by bacteria and fungi and are in turn an integral component of soil structure. After death they leave important nitrogenous waste. Because of their (regulating) role in decomposition and nutrient cycling as well as soil structure formation, their abundance, species composition and diversity in a particular habitat serve as good indicators of soil 'health' (Minor, 2004). This may apply to the larger groups of soil microarthropods.

SAMPLING METHODS

The protocols described here are to address primarily mites and collembolans present in the mineral horizons and litter. The principle is to take small cores or blocks and combine them to make a single, bulk sample, which is then subsampled for extraction by the Berlese–Tullgren method. Alternatively three or four composite samples are collected, instead of just one. Soil samples with (or without) litter are taken in the field with a split corer. The corer measures 3.5 by 3.5cm (width) and 10cm (height) and samples are taken to 5cm depth and then transferred to a plastic box (300mL volume, Figures 4.1a and b) to be stored in larger containers for transportation to the laboratory. Alternatively, soil samples can be collected from 3.5×3.5×5cm (depth) by using a small spade which turns a constant volume, ca. 100mL. Soil from 12 subsamples is then placed in a 5kg plastic bag as a composite.

The composite soil sample is then divided into ten unsealed plastic bags, averaging 1kg per bag, labelled, and all bags are transferred into a cloth sack, with the dimensions of 30×35cm, to prevent the death of mesofauna. Cloth sacks have been found very efficient for temporarily storing and transporting samples over longer distances and times because the cloth allows air to circulate and prevents unwanted rises in temperature. A plastic sack should not be used since it could substantially increase the soil temperature and may lead to death of the mesofauna before extraction. Also, unnecessary exposure to direct sunlight must be avoided during sample bagging in the field. Samples must also be protected from rain during transport from the field since animals could also be killed if the sample is too wet. Upon arrival at the base-camp or laboratory, all cloth bags are arranged under cover according to their sample sites. Soil samples are then packed in a carton pre-layered with paper to maintain moisture stability inside the box. Samples are then ready to be transported, preferably using an air-conditioned vehicle, to the laboratory for identification. Samples should be away from the heat of the vehicle's engine and exhaust system, and if the sample is transported with a non-air-conditioned vehicle, it should be placed on a passenger's seat. Both sunlight and engine heat are injurious to animals in the sample. Upon arrival in the laboratory, the soil is immediately submitted to extraction.

Figure 4.1 *Metal corer for sampling mesofauna from the litter and mineral soil; a) the content of each core is expelled into a labelled plastic container with a small trowel or field knife; b) gloves are a precaution against plant vines, ants, arachids and other arthropods*

EXTRACTION

Berlese–Tullgren

The Berlese–Tullgren apparatus (Figure 4.2) is used to extract the active, free-living mesofauna from the soil and litter samples. The apparatus is kept in a base camp laboratory near or in the benchmark area. The ideal set-up consists of a wooden container (cupboard) of about 160×50cm, divided into upper and lower compartments, with each compartment containing a frame to hold the plastic funnels (Figure 4.2a). The cabinet is fitted with doors to exclude flying insects not part of the sample. A sieve (of 8cm diameter and 5cm high), containing soil and litter samples is seated in the top of the funnel. The mesh size of the sieve is 1.5mm, but should contain four holes of 4mm diameter to allow for the escape of larger animals downwards. Funnels are gently heated by electric light bulbs (25 watts is ideal; if not available, 40 watts will do), suspended 14cm above the sieve. The heating should be increased gradually by increasing each day the length of time the bulbs are switched on, to prevent the sample material drying too fast which can immobilize the mesofauna before they escape, for up to about eight days. The basic principle of Berlese–Tullgren apparatus is that soil organisms will respond to the humidity reduction caused by increased temperature in the litter or soil by migrating downwards until they finally fall through the gauze (wire mesh) into a receiver container attached to the base of the funnel and containing preservative. Thus for greatest effectiveness, the funnel should have a steep slope and a smooth surface such that animals falling through the gauze slip immediately into the receiver bottle. The position of the lamp is adjustable, which provides for another way to ensure the temperature of the soil is raised gradually, thus preventing the slower moving species from becoming trapped in

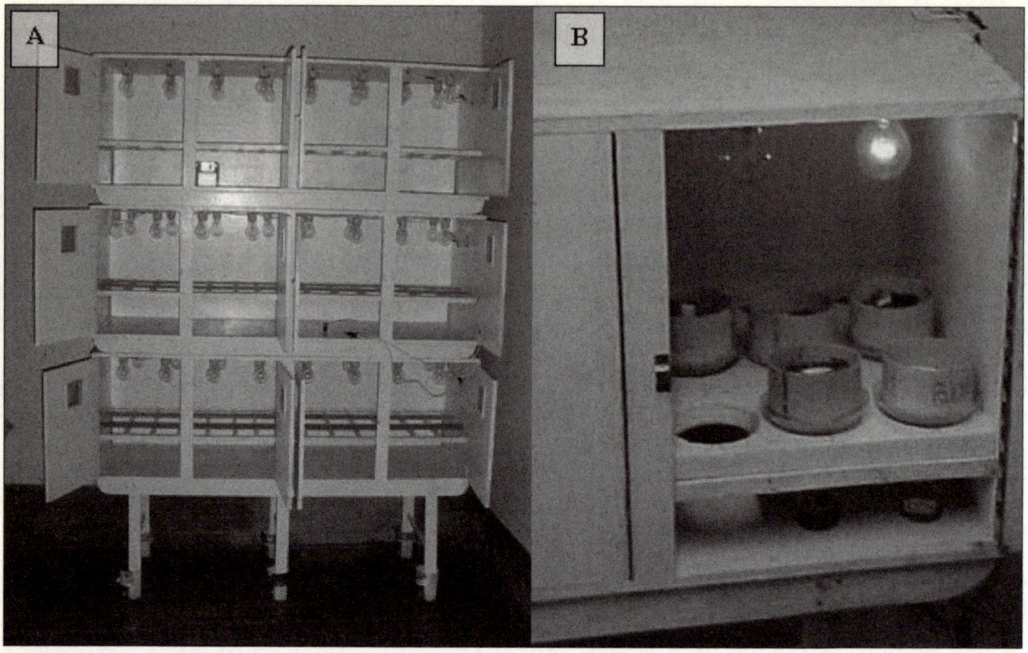

Note: Note that the legs stand in water tubs to prevent invasion by ants and other insects.

Figure 4.2 *a) Supporting container for Berlese–Tullgren funnels; b) a Berlese–Tullgren system in operation*

hard dry cakes of soil. Typically, the temperature of the sample is gradually increased from 27°C to approximately 40–45°C. The samples remain in the apparatus until they are completely dry (i.e. constant dry weight), which is usually within eight days. If left for a longer time, eggs will hatch and immature forms will become adults, so that the results will be less representative. The preservative is usually formalin at 4 per cent or 96 per cent ethanol.

Berlese method modified

When electricity is unavailable, the funnel can be set without light and the heating/drying process is carried out simply using ambient room temperature. A Berlese modified aluminium funnel, fitted with a 2×2mm (13 mesh) sieve is shown in Figure 4.3 and is suitable for unassisted drying. A Berlese system of this kind operated without light often gives a better extraction although it takes more time; with this modification the incubation time is 7 to 14 days, depending on initial soil moisture conditions. The receiving vial is filled with 96 per cent ethanol for preservation and harvested at four- to five-day intervals (Figure 4.3). The Berlese apparatus *sensu stricto* can be transformed to the Berlese–Tullgren model by fitting a light bulb of 10–40 watts above the soil samples. The lid on top is closed to prevent entry of other insects. With the presence of light, incubation time (amount of time needed to extract the faunas from soil) is approximately four to five days. The use of a light bulb might attract nocturnal insects, especially if the Berlese funnel is not properly closed. To avoid this, the ventilation hole on the cover should be closed at night.

Source: Courtesy of Dr Y. R. Suhardjono, LIPI Bogor

Figure 4.3 *Berlese funnels filled with soil and litters*

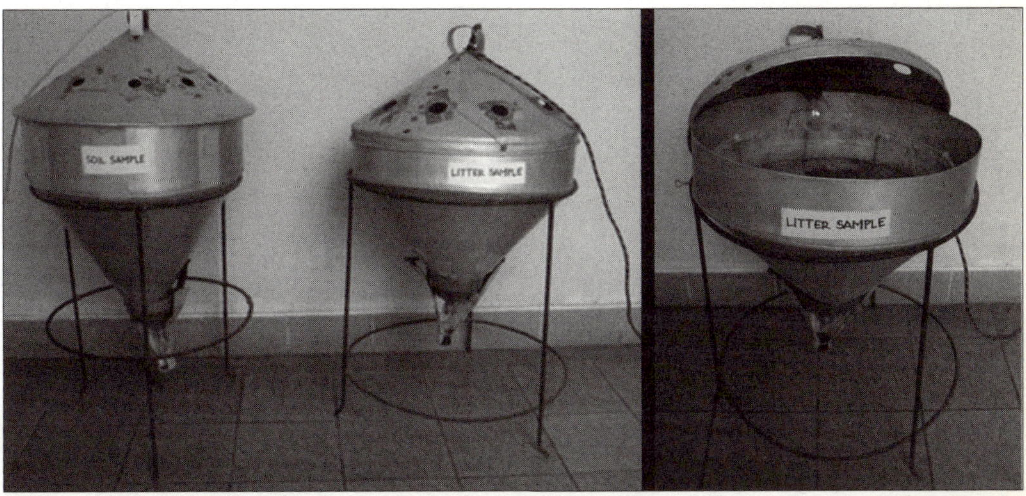

Note: Incubation takes place with the lid closed, but ventilation holes allow air in. A cloth fitted to the lid allows extraneous flying insects to be excluded from the holes. Collecting vials should be securely taped to the bottom of the funnels.

Source: courtesy of Dr Y. R. Suhardjono, LIPI Bogor

Figure 4.4 *A modified Berlese funnel made from aluminium; 50cm in total height with (top) funnel 40cm diameter*

Figure 4.5 *Basic equipment required to perform core-by-core Berlese–Tullgren extractions*

In spite of the lower efficiency compared to other sampling methods, the Berlese–Tullgren system, and to some extent the Berlese system, is one of the most frequently used extraction methods for assessing diversity and density of soil microarthropods (André et al, 2002). A useful feature of the systems is that they can be operated at a range of scales, and the same principle can be used to extract a single core or a large bulked sample. All that is required is a larger funnel and a longer drying time. They are also cheaper and simple to build, can be easily moved to the place of sampling and can be adjusted for large numbers of samples, even in remote regions (Franklin and Morais, 2006). Also the complete range of equipment required is inexpensive and can be easily improvised (Figure 4.5).

SPECIMEN CLEARING AND MOUNTING PROCESSES

For identification, mesofauna specimens are mounted on slides after clearing. The mounting process should be done under a dissecting microscope. Observation of morphology usually requires a compound microscope.

Clearing

Clearing is the process of making the tissues of the specimen more transparent by removing or washing away body fat and pigments. Nesbitt's solution (composition given below) is commonly used, but potassium hydroxide (KOH) and lactic acid are effective alternatives. Specimens can be cleared with 15 per cent KOH for two to five minutes, with or without heating, while Greenslade et al (2008) have proposed the use of 10 per cent KOH for one to five minutes depending on the pigment and body fat of the animal. After immersion in KOH, specimens are transferred into chloral phenol for a few minutes until the neutralizing reaction is complete. A simpler clearing method is the use of lactic acid, where mesofauna specimens are immersed and heated for periods of time depending on size. This method is cheaper, easier, safer and works well (Greenslade et al, 2008). Taxa with delicate cuticles, like some Diplura, Protura and Pauropoda, should not be placed in full-strength acid without proper care. Preferably the specimens should be placed in a 50 per cent lactic acid/ethanol mixture for a period of hours, without heating. To show special features like trichobotrium and hidden structures of the genitalia, the heating time must be controlled for each case, or the specimens may be destroyed or disarticulated. For permanent storage, specimens must be transferred to another medium, because lactic acid continues to clear and soften specimens. After identification, most oribatid mites are kept in labelled vials of 70 per cent alcohol to which 5 per cent glycerine is added. Mites can also be cleared and preserved from alcohol into Hoyer's medium. Highly sclerotized mites can be cleared with lactophenol (contains lactic acid, phenol crystals and distilled water in proportions of 50:25:25) or with lactic acid only.

Mounting

No methods are entirely satisfactory for mounting (Greenslade et al, 2008). The positioning of specimens on the glass is different for different orders: for example, in collembolans, Poduromorpha are placed dorsoventral, while Entomobryomorpha, Symphypleona and Neelipleona are placed latero-lateral.

Some groups need to be dissected before mounting for identification purpose to see the chaetotaxy of each part, commonly used as the key characteristics. Again using the collembolan example, dissection is recommended for Symphypleona and the larger (6–8mm) Paronellidae and Entomobryidae. Dissecting is addressed to various body parts such as head, legs, furcula, ventral tube and abdominal segments V and VI. Dissected body parts of Entomobryomorpha should be placed dorso-ventrally on the glass. Meanwhile, body parts of large Symphypleona are mounted as follows: head (dorso-ventral), big abdomen (latero-lateral), small abdomen and furcula (dorso-ventral). In the case of mutilated specimens, it is advisable to mount all body parts on one slide to avoid misidentification.

For mounting, Berlese solution is very suitable. Specimens are mounted on slides, covered with a cover slip and dried in the oven at 70°C for at least seven days. Other solutions, such as Hoyer's medium, can also be used for the mounting process (Table 4.1). Temporary mounting is commonly used for soil mites, and can also be used for other mesofauna groups. The specimen is transferred to a few drops of lactic acid or glycerine on a cavity slide. A cover slip is placed partly over the depression. The specimen is pushed under the cover slip near the edge of the depression. The specimen can be rolled gently beneath the cover slip to any desired orientation. Travè (1965), Krantz (1978) and Norton (1990) provide a good background discussion on clearing and mounting mites.

Table 4.1 *The chemical composition for clearing and mounting specimens*

1	Clearing specimens	
	Nesbitt's fluid	25mL distilled water, chloral hydrate, and 2.5mL hydrochloric acid.
	10% KOH or NaOH	
2	Mounting specimens	
	Berlese's mountant	20mL distilled water, 15g gum arabic, 50g chloral hydrates, 5mL glycerine, 10g glucose syrup, and 5mL glacial acetic acid
	Hoyer's medium	distilled water 50mL, gum arabic 30g, chloral hydrate 200g, and glycerine 20mL

Note: Berlese mountant: harden at 70°C for seven days. Hoyer's is a temporary agent and does not harden on heating, but preparations can be made semi-permanent by sealing the cover slip with clear nail varnish.

Specimen identification and data analyses

Identification

Most soil invertebrates can be identified to order or family with the aid of standard reference works, such as Bachelier (1978) and Dindal (1990). For most soil arthropod groups, however, identification beyond this level is very difficult. Taxonomic expertise is perhaps the major critical factor in selecting a taxon to study at the species level. Several pages are available on the web, providing keys that help with identification by pointing out distinctive characteristics of various invertebrate groups, for example www.cals.ncsu.edu/course/ent591k/kwikey1.html or http://soilbugs.massey.ac.nz/key.php. The latter also links to the main literature allowing identification at the generic and/or species level. Bachelier (1978) gives a good background within broad taxonomic units of soil arthropods, and includes illustrated keys for several groups. A comprehensive study of the biology, taxonomy and ecology of soil biotic groups was given by Dindal (1990). The book is written for easy use by both beginners and professionals, and contains illustrated identification keys at least to family level, and sometimes to species level, for several groups like Nematoda, Enchytraeidae, Scorpiones, Araneae, Pseudoscorpionida, Opiliones, Mites, Isopoda, Chilopoda, Pauropoda, Symphyla, Protura, Diplura, Collembola, Isoptera, Coleoptera, Diptera and Hymenoptera. Adis (2002) provides illustrated keys for beginners and anyone interested in Neotropical Arachnida and Myriapoda. The book has identification keys to all classes, orders, families, some genera, and lists of known terrestrial species, and also supplies references on the current taxonomy of these groups.

Suborders of mites and many families of collembolans are readily identified, once their basic morphology and characteristics are learned (Crossley and Coleman, 1999).

The basis of identification and classification is summarized in the references cited below. For instance, identification of collembolans to generic level is available in Greenslade et al (2008). Publications of Yoshii (1981a, 1981b, 1982a, 1982b and 1983) can be used to identify collembolans to species level in the families Entomobryidae and Paronellidae. The resources to identify mites to family level are available in Krantz (1978) and Dindal (1990). Balogh (1972), Balogh and Mahunka (1983) and Balogh and Balogh (1992) provide keys and figures to generic level for mites belonging to the suborder Oribatida. Unfortunately there are very few species-specific keys for this group in the tropics. Balogh and Balogh (1990) presents brief characterization and identification keys for oribatid mites inhabiting the Neotropical region, as well as a checklist and bibliography of described species from this area.

Data analyses

The number of individuals of each order from each site is used to estimate abundance. The samples made using small cores or blocks are combined to make a single bulk sample which is then subsampled for extraction using the Berlese–Tullgren method. This procedure overcomes the relatively low efficiency of the extraction, because it maximizes the chances of capturing representative groups and/or species. The data set should therefore include the species richness in those taxa selected for species-level identification. Given that groups of sites with similar features have a similar soil fauna, and that most soil animals can be allocated to basic functional groups (Ruf et al, 2003), poorly known taxa can also be classified at higher taxonomic levels only (order or family) or according to their major ecological function (detritivores, phytophagous, predators, etc.). It has been shown that data sets built at the genus and family systematic rank (for example, for oribatid mites) can also detect the effects of disturbance with little loss of information and can be a preliminary tool for describing patterns and successions in human-disturbed soil ecosystems (Caruso and Migliorini, 2006).

Comparison of the taxon composition between land uses should be performed by hierarchical cluster analysis using average linkage of agglomerative clustering based on transformed abundance. The number of individuals is commonly log transformed: $x' = \log10 (x + 1)$ to avoid the risk of overemphasizing dominant species in the data analysis (Ludwig and Reynolds, 1988). The program DIVERS from Ecological Methodology Package for Windows, can be used to calculate species-richness indices. A new statistical estimator that takes into account undetected, shared species, based on statistical correspondence between singletons and other rare species combinations amongst the samples being compared (Chao–Sørensen Abundance-based Estimator) was developed by Chao et al (2005) and Chao et al (2006). This index is especially useful for hyperdiverse invertebrate taxa with numerous rare species, such as collembolans, oribatid mites and ants. This estimator and also other diversity indices can be calculated using the program EstimateS (Colwell, 2006). Further analyses can be carried out, for example rarefaction, and multivariate statistics.

References

Adis, J. (2002) *Amazonian Arachnida and Myriapoda*, Pensoft, Sofia-Moscow

André, H. M., Ducarme, X. and Lebrun, Ph. (2002) 'Soil biodiversity: Myth, reality or conning?', *Oikos*, vol 96, pp3–24

Bachelier, G. (1978) 'La faune des sols, son écologie et son action', *Initiations et Documents Techniques*, no 38, ORSTOM, Paris

Balogh, J. (1972) *The Oribatid Genera of the World*, Akadémiai Kiadó, Budapest

Balogh, J. and Balogh, P. (1990) 'Oribatid mites of the Neotropical Region II', in J. Balogh and S. Mahunka (eds) *The Soil Mites of the World*, Elsevier, Amsterdam

Balogh, J. and Balogh, P. (1992) *The Oribatid Mite Genera of the World*, vols 1 and 2, Hungarian Natural History Museum Press, Budapest

Balogh, J. and Mahunka, S. (1983) *Primitive Oribatids of the Palaearctic Region*, Elsevier, Amsterdam

Caruso, T. and Migliorini, M. (2006) 'Micro-arthropod communities under human disturbance: Is taxonomic aggregation a valuable tool for detecting multivariate change? Evidence from Mediterranean soil oribatid coenoses', *Acta Oecologica*, vol 30, no 1, pp46–53

Chao, A., Chazdon, R. L., Colwell, R. K. and Shen, T-J. (2005) 'A new statistical approach for assessing compositional similarity based on incidence and abundance data', *Ecology Letters*, vol 8, no 2, pp148–159

Chao, A., Chazdon, R. L., Colwell, R. K. and Shen, T-J. (2006) 'Abundance-based similarity indices and their estimation when there are unseen species in samples', *Biometrics*, vol 62, no 2, pp361–371

Colwell, R. K. (2006) 'EstimateS: Statistical estimation of species richness and shared species from sample; Version 8 User's Guide and Application', http://viceroy.eeb.uconn.edu/estimates.

Crossley, D. and Coleman, D. (1999) 'Microarthropods', in M. E. Sumner (ed) *Handbook of Soil Science*, CRC Press, Boca Raton, pp59–65

Dindal, D. L. (1990) *Soil Biology Guide*, John Wiley & Sons, New York

Franklin, E. and Morais, J. W. (2006) 'Soil mesofauna in Central Amazon', in F. M. S. Moreira, J. O. Siqueira and L. Brussaard (eds) *Soil Biodiversity in Amazonian and Other Brazilian Ecosystems*, CABI Publishing, Wallingford, pp142–162

Greenslade, P., Deharveng, L., Bedos, A. and Suhardjono, Y. R. (2008) *Handbook to Collembola of Indonesia*, Fauna Malesiana, Brill, Leiden (in press)

Krantz, G. W. (1978) *A Manual of Acarology*, Oregon State University Book Store Inc, Oregon

Ludwig, J. A. and Reynolds, J. F. (1988) *Statistical Ecology: A Primer on Methods and Computing*, John Wiley and Sons, New York

Minor, M. (2004) 'Soil mites and other animals', www.massey.ac.nz/~maminor/mites.html, accessed 10 January 2008

Norton, R. A. (1990) 'Acarina: Oribatida', in D. L. Dindal (ed) *Soil Biology Guide*, John Wiley and Sons, New York, pp779–803

Ruf, A., Beck, L., Dreher, P., Hund-Rinke, K., Römbke, J. and Spelda, J. (2003) 'A biological classification concept for the assessment of soil quality: "Biological soil classification scheme" (BBSK)', *Agriculture, Ecosystems and Environment*, vol 98, nos 1–3, pp263–271

Siepel, H. (1990) 'Niche relationships between two panphytophagous soil mites, *Nothrus silvestris* Nicolet (Acari, Oribatida, Nothridae) and *Platynothrus peltifer* (Koch) (Acari, Oribatida, Camisiidae)', *Biology and Fertility of Soils*, vol 9, no 2, pp139–144

Travè, J. (1965) 'Quelques techniques de récolte, de triage, d'observation et de conservation des Oribates (Acariens) et autres microathropodes', *Revue d'Écologie et de Biologie du Sol*, vol 2, pp23–47

Yoshii, R. (1981a) 'Paronellid Collembola of Sabah', Entomological Report from the Sabah Forest Research Centre No. 3, Japan International Cooperation Agency, Tokyo, pp1–70

Yoshii, R. (1981b) 'Neanurid Collembola of Sabah', Entomological Report from the Sabah Forest Research Centre No. 4, Japan International Cooperation Agency, Tokyo, pp1–70

Yoshii, R. (1982a) 'Lepidocyrtid Collembola of Sabah', Entomological Report from the Sabah Forest Research Centre No. 5, Japan International Cooperation Agency, Tokyo, pp1–38

Yoshii, R. (1982b) 'Studies on the Collembolan Genera *Callyntrura* and *Dicranocentroides*', *Entomological Report from the Sabah Forest Research Centre No.* 6, Japan International Cooperation Agency, Tokyo, pp1–38

Yoshii, R. (1983) 'Studies on Paronellid Collembola of East Asia', *Entomological Report from the Sabah Forest Research Centre No.* 7, Japan International Cooperation Agency, pp1–28

Chapter 5

Soil Nematodes

Juvenil E. Cares and Shiou P. Huang (deceased)

Introduction

Nematodes are mostly small, water-dependent invertebrates considered to be the most abundant, and one of the most diverse, of any group of animals. Due their capacity for adaptation, nematodes are present anywhere organic carbon is available, at all latitudes of the planet, and from the bottom of the sea to the tops of the mountains. The word nematode is derived from the Greek 'nema', meaning in the shape of a thread, since they possess an elongate and cylindrical body. Evolution adapted nematodes to explore a variety of food sources. The parasites colonize and feed in plant or animal tissues, while the free-living forms are grazers or predators of small organisms, such as bacteria, protozoa, fungi, algae and micro-invertebrates.

Nematodes are primarily known as a threat to human, animal and plant health. Nevertheless, they play equally important roles as agents of nutrient cycling, and as regulators of soil fertility through both energy flow and nutrient mobilization and utilization (Procter, 1990). Although nematode biomass in soil is small (10^1 to 10^4mg dry wt m^{-2}), they are responsible for 10 to 15 per cent of the soil animal respiration (Sohlenius, 1980; Petersen, 1982). Besides these ecosystem benefits, nematodes can also provide important agroecosystem services as agents of biological control of insect pests, for example entomophilic nematodes such as *Deladenus siricidicola* and entomopathogenic nematodes such as *Steinernema carpocapsae* and *Heterorhabditis bacteriophora*. Their omnipresence and abundance in all types of soil and water habitats, as well as their short life cycle and sensitivity to environmental alterations, qualify them as powerful bioindicators of ecological conditions (Huang and Cares, 2006).

Soil nematodes live in assemblages usually composed of five major functional groups: plant parasites, bacterial feeders, fungal feeders, predators and omnivores (Plate 2). Therefore nematodes are well represented throughout the soil food web. Although full taxonomic identification of nematodes requires intensive training, in most cases identification at the level of functional groups can be readily accomplished on the basis of morphology of the feeding apparatus. In plant-parasitic nematodes the feeding apparatus includes a needle-like stylet (Plate 2a); plant-parasitic nematodes may cause significant yield losses in several crops. Fungal feeders, mostly equipped with a small and delicate stylet (Plate 2b), feed on hyphae of saprophytic, pathogenic, beneficial and mycorrhizal fungi. Some fungivorous nematodes are facultative plant parasites. Bacterial feeders, without the stylet (Plate 2c), can regulate the available nitrogen and phosphorus for plants, influence *Rhizobium* nodulation, and consume and disseminate beneficial and plant pathogenic bacteria. As typical r-strategists, bacterial feeding nematodes may see their populations increase under soil disturbance or nutrient enrichment conditions, this being interpreted as an indicator of soil fertility (Ferris et al, 2001). Predators have their mouth cavity equipped with one or more tooth-like structures (Plate 2d), or in some cases one stylet, and prey on other nematodes, microinvertebrates and protozoa, and in some cases they eat bacteria and the spores of fungi. Omnivores have their mouth cavity armed with a hollow stylet (Plate 2e); they are polyphagous, feeding in all trophic capacities, as fungal and bacterial feeders, herbivores, and also as predators eating other nematodes.

Some nematodes may occur in a wide range of habitats, whereas others are more restricted. They also differ among themselves in sensitivity to soil disturbances and to chemical pollutants. As proposed by Bongers (1990) and Bongers and Bongers (1998), colonizer nematodes (similar to r-strategists) typically possess a short generation time, produce many small eggs, show the presence of dauerlarvae, and increase their populations under food-rich conditions, being more resistant to soil disturbances and to soil pollutants. On the other hand, persister nematodes (similar to K-strategists) are characterized by a long generation time, production of fewer but larger eggs, low motility, the absence of dauerlarvae and sensitivity to pollutants and other disturbance factors. Therefore, appropriate analyses of the nematode community, taking in account the diversity, abundance and community structure, may present an index of land use systems and land use change, and also measure soil disturbance levels due to pollutants or other factors.

Soil sampling

At each sampling point in the grid system (see Chapter 2), two circles, of 3m and 6m radius, are marked out. A steel soil corer is used to make four equidistant samplings on the smaller circle, and eight equidistant cores on the larger one, all cores being to 20cm depth. The 12 soil cores are bulked into a plastic bag, which should be sealed to avoid desiccation, and also kept out of sunlight. Samples are transported in an insulated box to the laboratory, and stored at 4°C until extraction; however, this should be done as soon as possible.

Nematode extraction

The soil sample is thoroughly mixed and a 300mL subsample is added to two litres of water in a bucket, agitated for 30 seconds, and left for another two minutes for soil particles to sediment. The suspension is then passed through a 45 to 60 mesh (250 to 350μm) screen, and nematodes are collected directly on a second screen of 400 mesh

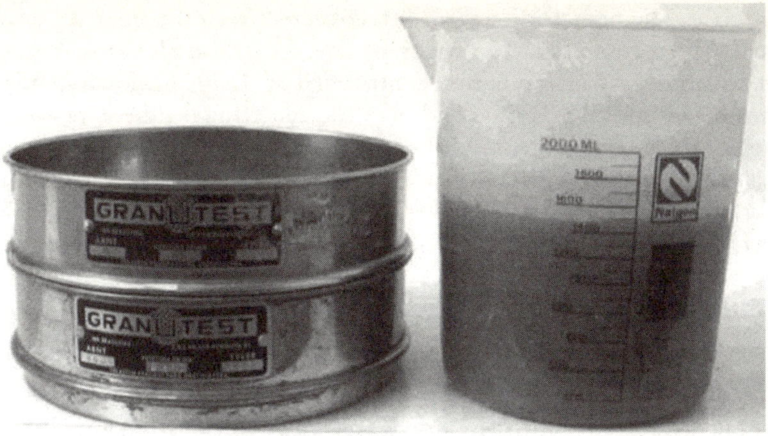

Figure 5.1 *At left, a set of metal sieves (45 mesh on the top and 400 mesh on the bottom), and right, the sedimenting soil suspension*

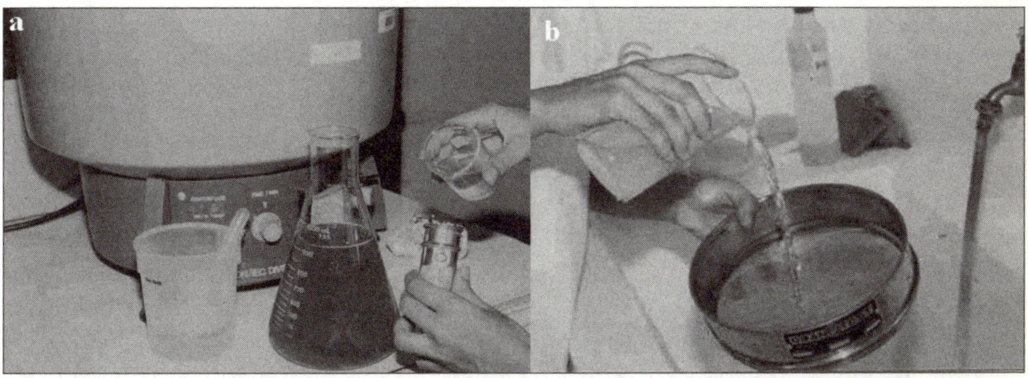

Figure 5.2 *a) Adding sucrose solution to re-suspend the soil and nematode mixture deposited in the bottom of the centrifuge tube; b) sieving of nematode suspension after sugar flotation*

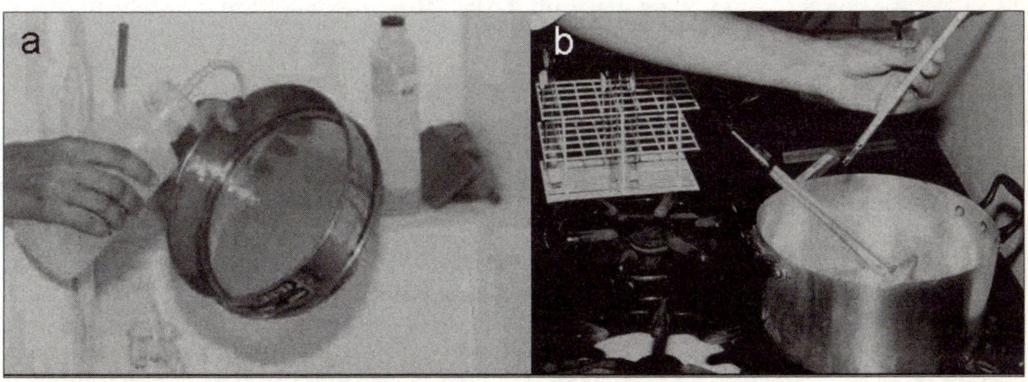

Figure 5.3 *a) Nematodes recovered by backwashing from the 400 mesh screen; b) killing of nematodes in hot water. The rack of test tubes shown is immersed in the pan of water*

(37μm) (Figure 5.1). The nematode suspension collected from the final screen is further clarified by a modified centrifugation sugar flotation method (Jenkins, 1964). In this procedure, an aqueous suspension is first centrifuged at 3500rpm for five minutes and the supernatant discarded. Next, the residue in the centrifugal tube is re-suspended in sucrose solution (456g L^{-1}) (Figure 5.2a), and re-centrifuged at 1000rpm for one to two minutes; nematodes are then collected from the supernatant by a 37μm screen (Figure 5.2b). Specimens on the screen are washed with clean water, and collected in a beaker by backwashing (Figure 5.3a). From the beaker the suspension is dispensed into test tubes for fixation.

Nematode fixation and counting

The extracted nematodes are killed by hot water at 50–60°C for a maximum of one minute (Figure 5.3b), and then fixed with Golden solution 2X (Hooper, 1970). After heating, the tubes are allowed to stand for 30 minutes and excess of supernatant is removed by pipette or vacuum suction (preferred), leaving 10–20mL of nematode suspension as a residue. An equal volume of Golden solution 2X is then added, giving a final formalin concentration of 3.2 per cent. The suspension is adjusted to 15mL and the total nematode population is counted by randomly removing 1mL aliquots, the total number being the mean of three subsample counts ×15.

Glycerine infiltration and nematode observation

Seinhorst's method (Seinhorst, 1959) is modified to avoid the laborious task of picking up nematodes one by one to make large numbers of permanent slides. In the modification, the nematode suspension is reduced to 3mL (Figure 5.4), then 7mL of Seinhorst I solution are added (Figure 5.5, Table 5.1) and the whole poured into a 5cm diameter Petri dish, and then placed in a desiccator at 43°C overnight (Figure 5.6). The suspension, being removed from the desiccator on the second day, is dried at the same temperature for four hours or more, with the lid of the Petri dish removed, to reduce at least half the volume. After being made up to the same volume (10mL) with Seinhorst II solution, the dish is closed, returned to the desiccator and left overnight again. The process is repeated three times; on the last occasion it is not necessary to add Seinhorst II, but the incubation is maintained at same temperature for a further 48 hours, to evaporate all alcohol. The procedure is illustrated in Figure 5.5; the composition of the solutions is given in Table 5.1.

After this process, the nematodes from the dish are mounted on slides in an anhydrous glycerine drop. Before placing the cover glass, a strip of adhesive tape is placed along one side of the slide, to prevent the cover glass crushing the specimens, and the whole preparation finally sealed with Canada balsam (Huang et al, 1984), (Figure 5.7). One hundred nematodes from the slides prepared from each sample are randomly selected for identification to genus level by using a compound light microscope (400–1000× magnification is needed). Nematode identification by morphological features is based on a specialized taxonomic literature (Goodey, 1951; Thorne, 1961; Goseco et al, 1974; Andrássy, 1983; Bongers, 1988, 1994; Smart and Nguyen, 1988;

Figure 5.4 *Removing excess water from the test tubes without disturbing the bottom part of the nematode suspension*

Anderson and Potter, 1991; Baldwin and Mundo-Ocampo, 1991; Decraemer, 1991; Fortuner, 1991; Geraert, 1991; Loof, 1991; Maggenti, 1991; Nickle and Hooper, 1991; Raski, 1991; Smart and Nguyen, 1991; Jairajpuri and Ahmad, 1992; Hunt, 1993; May et al, 1996; Cares and Huang, 2000, 2001).

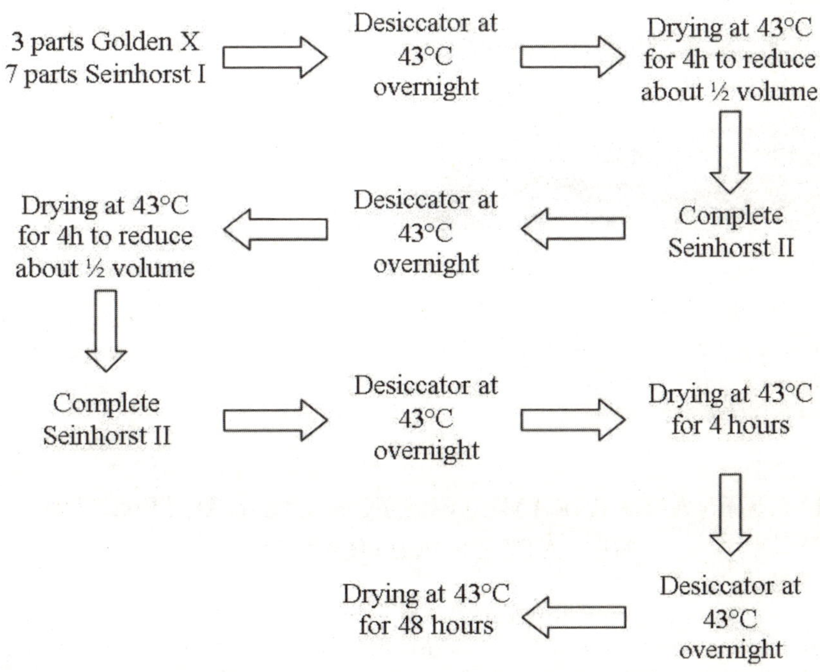

Figure 5.5 *Scheme of Modified Seinhorst's method (process of infiltration of glycerine) for a bulk nematode population. Petri dishes should not contain more than one-third fluid*

Table 5.1 *Composition of reagents used for nematode fixation and glycerine infiltration*

	Golden X	*Golden 2X*
Formalin (+ 40% formaldehyde)	8 parts	16 parts
Glycerine	2 parts	4 parts
Distilled water	90 parts	80 parts
	Seinhorst I	*Seinhorst II*
96% alcohol	20 parts	95 parts
Glycerine	2 parts	5 parts
Distilled water	78 parts	

Note: The bottom of the container is maintained at 43°C.
Source: Photo Marcella A. Teixeira

Figure 5.6 *Desiccator containing Petri dishes with nematodes for glycerine infiltration*

Figure 5.7 *Tray with permanent mounts of nematode specimens*

Indices and parameters for characterizing nematode populations

Nematodes identified to genus level can be allocated into five trophic groups, namely bacterial and fungal feeders, plant parasites, omnivores and predators, based on the criteria of Yeates et al (1993). Data on absolute frequency, total abundance and relative abundance of each of the genera may be used for calculating the various diversity and evenness indices using the formulae as given below. Frequency and abundance data for the various trophic groups are used to calculate the trophic diversity index and various

Plate 1 *The Winkler system for extracting ants and beetles a) One square metre of litter is delimited by the square frame; b) removing litter working from the rim to the centre, wearing leather gloves; c) litter is placed into the sieve and sieved to separate the thicker fragments; d) sieved material is transferred to field bags; e) each sample of sieved litter is transferred into the net bag; f and g) net bags are fixed inside the Winkler extractor, already hanging in a well-aired space with cups basally fitted containing 70% alcohol.*

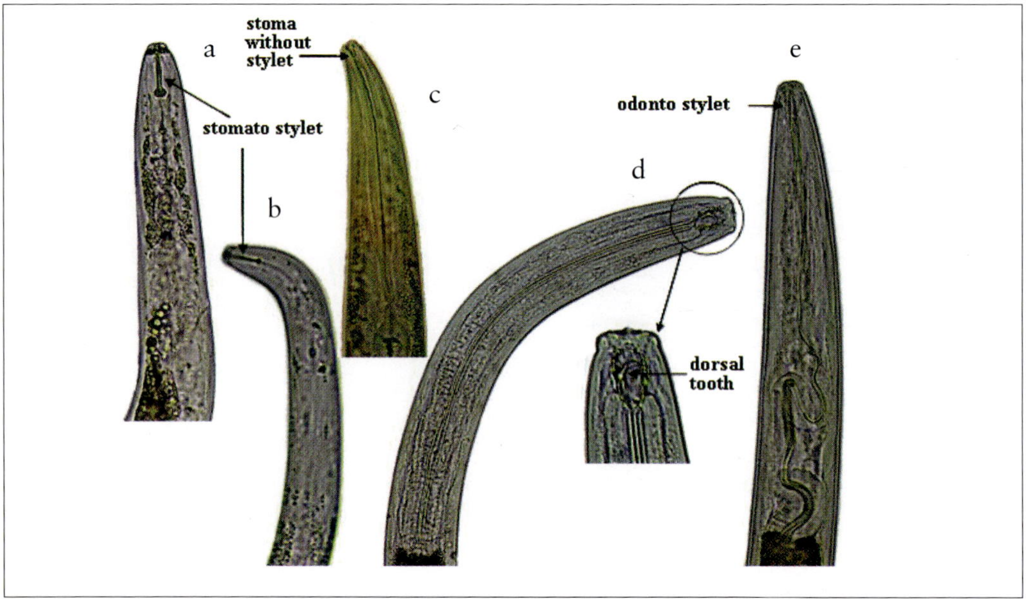

Plate 2 *Photomicrographs of anterior region of soil nematodes of different functional groups showing their feeding apparatus (arrows): a) plant parasite; b) fungal feeder; c) bacterial feeder; d) predator; e) omnivore.*

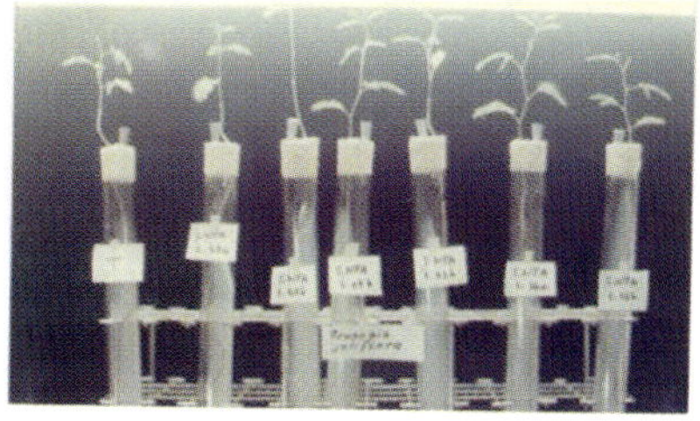

b) Agar slants with nutrient solution

a) Sirato in plastic pouches with nutrient solution

c) and e) Plant tests isolate efficiencies in Leonard jars

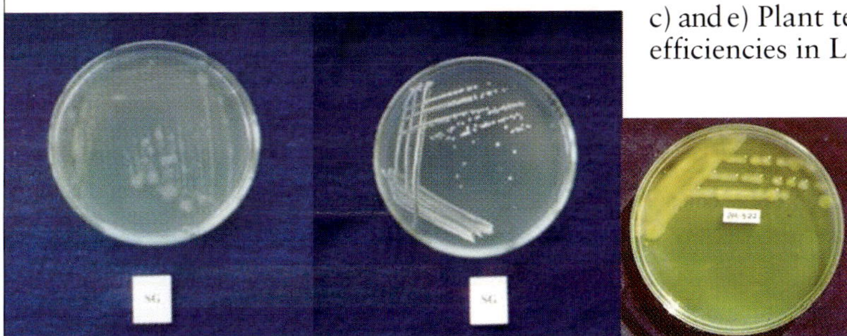

d) Characteristics of two *Bradyrhizobium isolates* (SG–slow growers) and *R. topici* (fast grower) grown on YMA

(i) (ii)

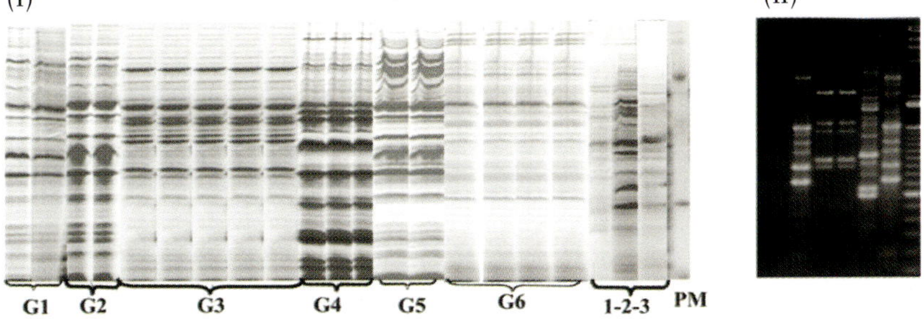

f) (i) SDS-PAGE total protein and (ii) Rep-PCR profiles of diverse strains

Plate 3 *Illustration of the steps in the methodology for nitrogen-fixing Leguminosae-nodulating bacteria (NFLNB) characterizations applied in the CSM-BGBD Project.*

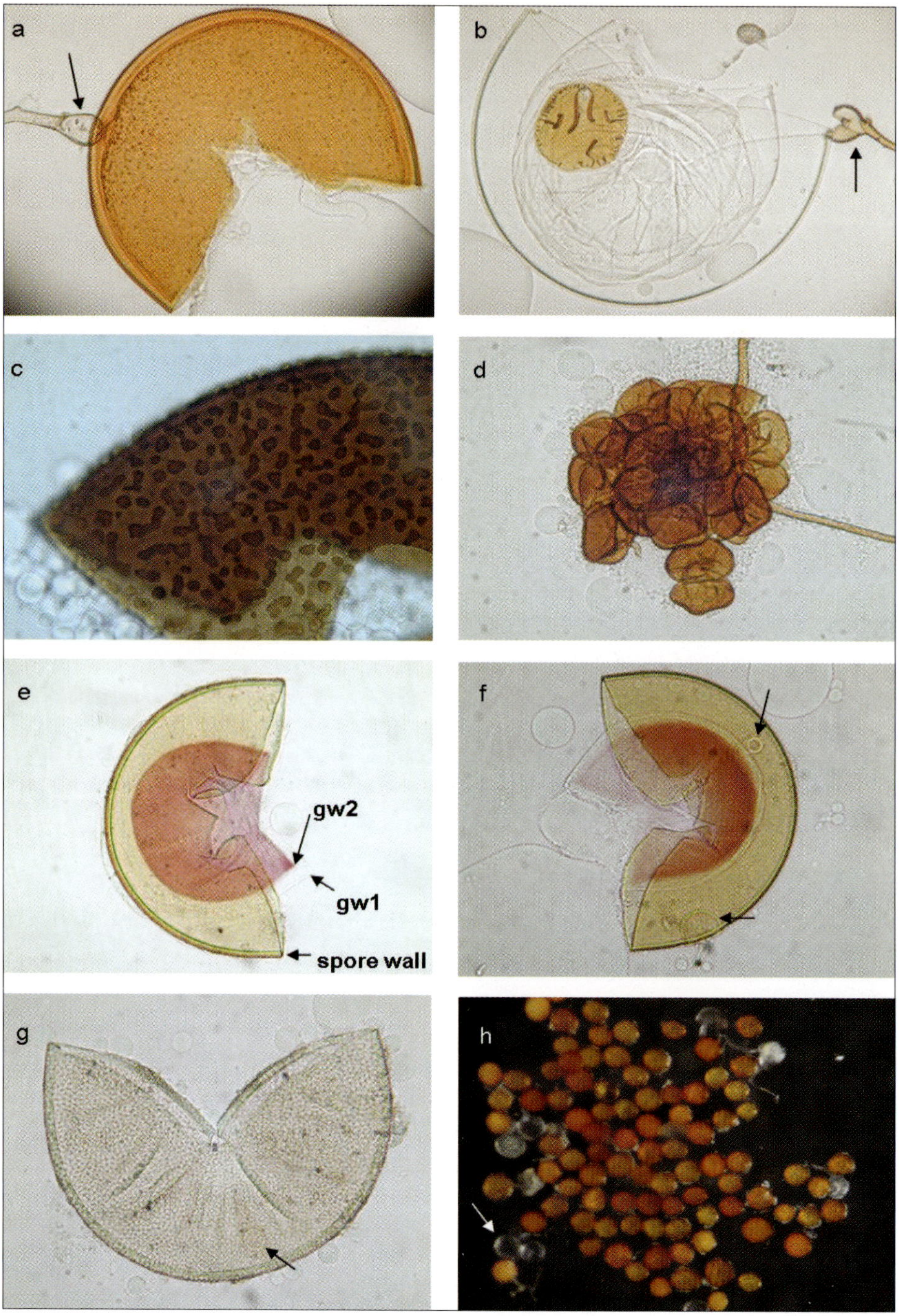

Plate 4 *Spores and structures produced by AMF species of the family Gigasporaceae: a) spore of* Gigaspora albida *showing the suspensor bulbous cell (arrow) typical of the family, b) spore of* Scutellospora scutata *with suspensor bulbous cell (arrow); notice round brown germination shield contrasting with the hyaline spore colour, c) detail of the spore wall ornamentation (warts) of* Scutellospora coralloidea, *d) knobby auxiliary cells differentiated by members of* Scutellospora. *Spores and structures produced by AMF species of the family Acaulosporaceae: e) spore of* Entrophospora colombiana *indicating the spore wall, germinal wall 1 (gw1) and germinal wall 2 (gw2) with its innermost layer reacting in Melzer's reagent, f) spore of* Entrophospora colombiana *showing the two scars (arrow), g) spore of* Acaulospora scrobiculata *showing the scar (arrow) left on the spore after the sporiferous saccule detached, h) spores of* Acaulospora *sp. showing some sporiferous saccules attached to the spores (arrow).*

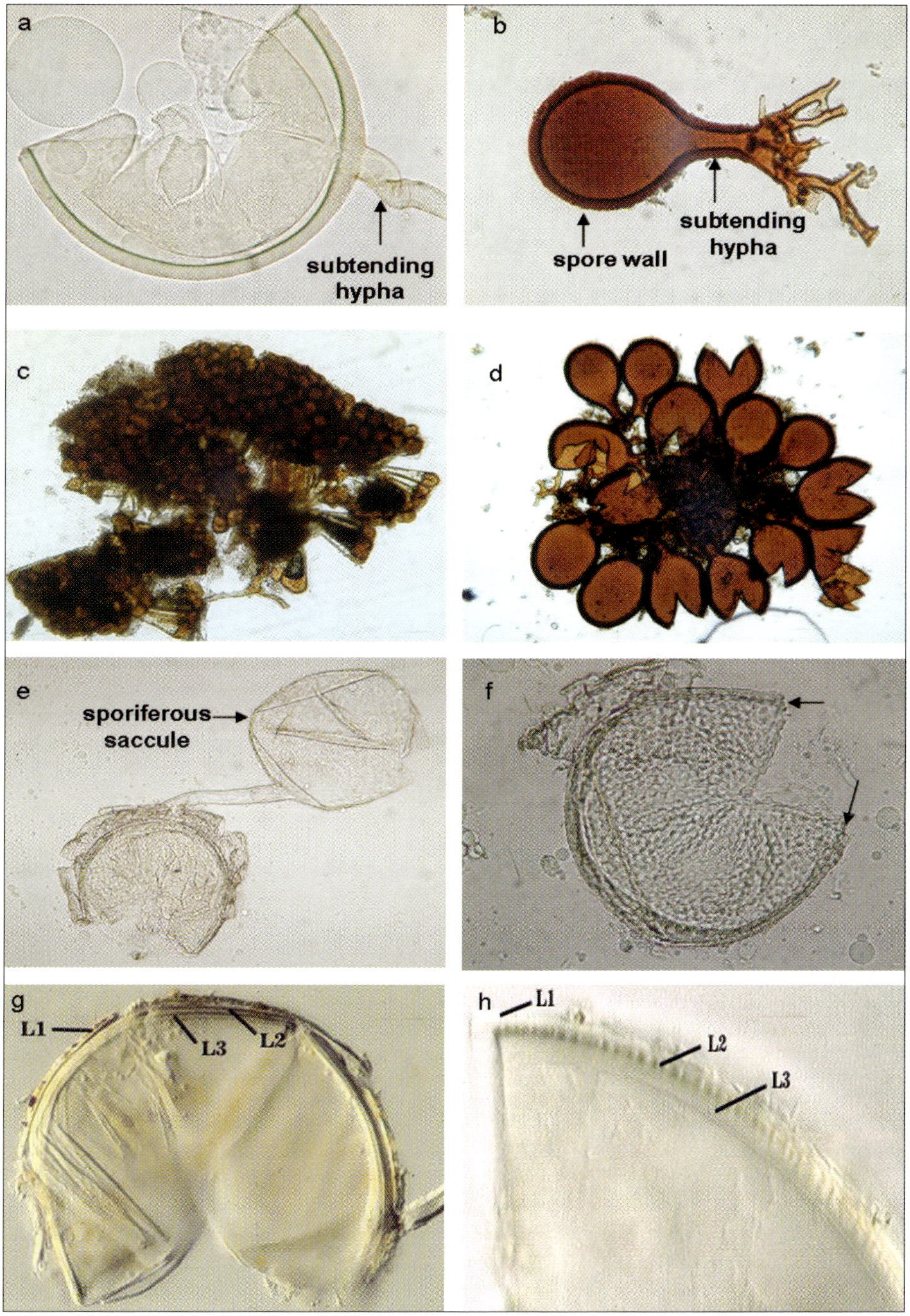

Plate 5 *Spores and structures produced by AMF species of the family Glomeraceae: a) spore of* Glomus clarum *indicating the subtending hypha; notice that the innermost layer of the spore wall detaches and looks similar to a germinal wall; b) spore of* Glomus *sp. showing the subtending hypha wall continuous with the spore wall; c) sporocarp of* Glomus clavispora, *d) sporocarp of* Glomus *sp.*
Spores and structures produced by AMF species of the family Archaeosporaceae (e and f) and Paraglomeraceae (g and h): e) spore of Archaeospora leptoticha *with sporiferous saccule; f) detail of* Archaeospora leptoticha *showing protuberances and depressions of layers 2 and 3 of the spore wall (arrows); g) spore of Paraglomus occultum showing the spore wall structure formed by three layers (L1, L2 and L3); h) spore of* Paraglomus brasilianum *showing the spore wall structure formed by three layers (L1, L2 and L3) (notice that L2 is ornamented with minute ridges). Pictures of* Paraglomus occultum *and* P. brasilianum *are from http://invam.caf.wvu.edu.*

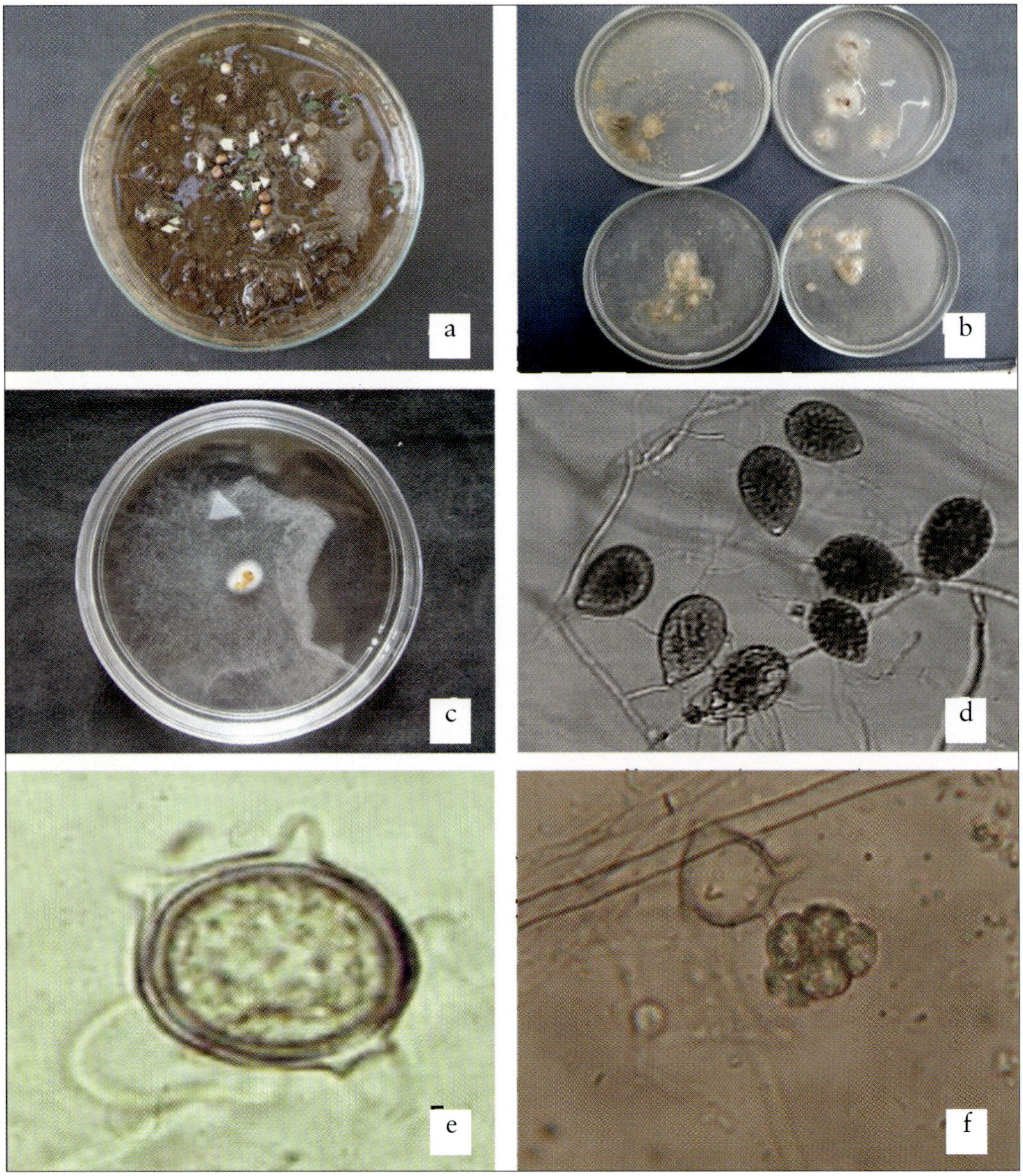

Plate 6 *Isolation of zoosporic fungi from environmental samples (Peronosporomycetes, Cytridiomycetes): a) soil sample with baits; b) water sample with baits; c) cure culture on bait; d) sporangia of* Phytophthora*; e) oogonium and antheridum of* Pythium*; f)* Pythium *liberating zoospores.*

Plate 7 *Soil washing technique for isolation of soil microfungi: a) pre-washing; b) sieves with different mesh; c) pre-washed soil; d) continued washing procedure; e) washed soil particles; f) plating on culture medium; g) plate with soil particles for incubation; h) fungal colonies on culture medium with growth retardants.*

Plate 8 *a) Larvae and pupae of fruit fly in the soil; b) MacPhail trap containing a food bait; c) common wing patterns with stripes; d) extrovertion of the aculeus; e) adult of fruit fly and detail of apical aculeus.*

ratios that consider the relative abundance of the trophic groups. If one nematode has two types of feeding habits, its population number is divided by two for each habit.

- Genus richness index [$d = (S - 1)/\log N$,

where S = number of genera, and N = total number of nematodes].

- Simpson's diversity index [$Ds = 1 - \Sigma(p_i)^2$,

where p_i = percentage of genus 'i' in the total abundance].

- Shannon's diversity index [$H' = -\Sigma\, p_i \log_2 p_i$].

- Evenness of Simpson's diversity index [$Es = Ds/Ds_{max}$,

where $Ds_{max} = 1 - 1/S$ and evenness of Shannon's diversity index ($J' = H'/H'_{max}$, where $H'_{max} = \text{Log}_2\, S$].

- Trophic diversity index [$T = 1/\Sigma(p_i)^2$,

where p_i = relative abundance of one trophic group; see Norton, 1978; Magurran, 1988; Krebs, 1994].

- The ratios of fungivores/bacterivores (FF/BF) and of (fungivores+bacterivores)/plant parasites [(FF+BF)/PP].

- Percentages of criconematids and of dorylaimids in the population.

Bongers (1990) allocated soil nematodes to a series from colonizers (c) to persisters (p) (similar to r- and K-strategists, respectively) by giving them a 'cp scale value' from 1 to 5. The cp1 colonizers are characterized by a short generation time, the production of many small eggs, always being active, the presence of dauerlarvae and growth under food-rich conditions. In contrast, the cp5 persisters are distinguished by long generation time, the production of few but large eggs, low motility, the absence of dauerlarvae and a high sensitivity to pollutants and other disturbance factors (Bongers and Bongers, 1998). Based on these concepts, several indices have been used as soil assessments: the maturity index (MI) (including only free-living nematodes) (Table 5.2) and the plant parasitic index (PPI) (including only plant parasites) (Table 5.3). To measure the soil disturbance level (Bongers, 1990), both indices are calculated by the same formula, $\Sigma(v_i \times f_i)$ (where $v_i = c - p$ value from 1 to 5 for genus 'i', and f_i = relative frequency of genus 'i'). Yeates (1994) expanded the maturity index to all soil nematodes (mMI). Bongers and Bongers (1998) proposed MI_{2-5} (same as MI, but with cp1 nematodes excluded) to evaluate pollution-induced stress factors, and the PPI/MI ratio as an indicator of soil fertility.

Table 5.2 *Families and cp values used for the maturity index**

Family name	cp value	Family name	cp value
Neotylenchidae	2	Achromadoridae	3
Anguinidae	2	Ethmolaimidae	3
Aphelenchidae	2	Cyatholaimidae	3
Aphelenchoididae	2	Desmodoridae	3
Rhabditidae	1	Microlaimidae	3
Alloionematidae	1	Odontolaimidae	3
Diploscapteridae	1	Aulolaimidae	3
Bunonematidae	1	Bastianiidae	3
Cephalobidae	2	Prismatolaimidae	3
Ostellidae	2	Ironidae	4
Panagrolaimidae	1	Tobrilidae	3
Myolaimidae	2	Onchulidae	3
Teratocephalidae	2	Tripylidae	3
Diplogasteridae	1	Alaimidae	4
Neodiplogasteridae	1	Bathyodontidae	4
Diplogasteroididae	1	Mononchidae	4
Tylopharyngidae	1	Anatonchidae	4
Odontopharyngidae	1	Nygolaimidae	5
Monhysteridae	1	Dorylaimidae	4
Xyalidae	2	Chrysonematidae	5
Linhomoeidae	3	Thornenematidae	5
Plectidae	2	Nordiidae	4
Leptolaimidae	3	Qudsianematidae	4
Halaphanolaimidae	3	Aporcelaimidae	5
Diplopeltidae	3	Belondiridae	5
Rhabdolaimidae	3	Actinolaimidae	5
Chromadoridae	3	Discolaimidae	5
Hypodontolaimidae	3	Leptonchidae	4
Choanolaimidae	4	Diphtherophoridae	3

Note: *after Bongers (1990).

Table 5.3 *Families and cp values used for plant parasitic index**

cp2	cp3	cp4	cp5
Tylenchidae	Dolichodoridae	Trichodoridae	Longidoridae
Psilenchidae	Hoplolaimidae		
Tylodoridae	Pratylenchidae		
Ecphyadophoridae	Heteroderidae		
Paratylenchidae	Hemicycliophoridae		
Anguinidae			

Note: *modified from Bongers (1990).

References

Anderson, R. V. and Potter, J. W. (1991) 'Stunt nematodes: *Tylenchorhynchus*, *Merlinius*, and related genera', in W. R. Nickle (ed) *Manual of Agricultural Nematology*, Marcell Dekker, New York, pp529–586

Andrássy, I. (1983) *A Taxonomic Review of the Suborder Rhabditina (Nematoda Secernentia)*, Eötvös Lorand University, Budapest

Baldwin, J. G. and Mundo-Ocampo, M. (1991) 'Heteroderinae cyst- and non-cyst-forming nematodes', in W. R. Nickle (ed) *Manual of Agricultural Nematology*, Marcel Dekker, New York, pp275–362

Bongers, T. (1988) *De Nematoden van Nederland*, Koninklijke Nederlandse Natuurhistorische Vereniging, Bibliotheekuitgave no 46, Pirola, Schoorl, The Netherlands

Bongers, T. (1990) 'The maturity index: An ecological measure of environmental disturbance based on nematode species composition', *Oecologia*, vol 83, pp14–19

Bongers, T. (1994) *De Nematoden van Nederland*, 2nd edition, Koninklijke Nederlandse Natuurhistorische Vereniging, Bibliotheekuitgave no 46, Pirola, Schoorl, The Netherlands

Bongers, T. and Bongers, M. (1998) 'Functional diversity of nematodes', *Applied Soil Ecology*, vol 10, no 3, pp239–251

Cares, J. E. and Huang, S. P. (2000) 'Taxonomia atual de fitonematóides, Chave sistemática simplificada para gêneros Parte I', *Revisão Anual de Patologia de Plantas*, vol 8, pp185–223

Cares, J. E. and Huang, S. P. (2001) 'Taxonomia de fitonematóides; Chave sistemática simplificada para gêneros Parte II', *Revisão Anual de Patologia de Plantas*, vol 9, pp177–235

Decraemer, W. (1991) 'Stubby root and virus vector nematodes: *Trichodorus*, *Paratrichodorus*, *Allotrichodorus*, and *Monotrichodorus*', in W. R. Nickle (ed) *Manual of Agricultural Nematology*, Marcel Dekker, New York, pp587–626

Ferris, H., Bongers, T. and Goede, R. G. M. (2001) 'A framework for soil food web diagnostics: Extension of the nematode faunal analysis concept', *Applied Soil Ecology*, vol 18, no 1, pp13–29

Fortuner, R. (1991) 'The Hoplolaimidae', in W. R. Nickle (ed) *Manual of Agricultural Nematology*, Marcel Dekker, New York, pp669–720

Geraert, E. (1991) 'Tylenchidae in agricultural soils', in W. R. Nickle (ed) *Manual of Agricultural Nematology*, Marcel Dekker, New York, pp795–826

Goodey, T. (1951) *Soil and Freshwater Nematodes*, Methuen and Co Ltd, London

Goseco, C. G., Ferris, V. R. and Ferris, J. M. (1974) *Revision in Leptonchoidea* (Nematoda: Dorylaimida), Purdue Nematode Collection, Department of Entomology, Purdue University, West Lafayette, IN

Hooper, D. F. (1970) 'Handling, fixing, staining and mounting nematodes', in J. F. Southey (ed) *Laboratory Methods for Work with Plant and Soil Nematodes*, Ministry of Agriculture, Fisheries and Food, Technical Bulletin No 2, pp34–39

Huang, S. P. and Cares, J. E. (2006) 'Nematode communities in soils under different land-use systems in Brazilian Amazon and savanna vegetation', in F. M. Moreira, J. O. Siqueira and L. Brussaard (eds) *Soil Biodiversity in Amazonian and other Brazilian Ecosystems,* 1st edition, CAB International, Wallingford, pp163–183

Huang, C. S. Bittencourt, C. and Silva, E. F. S. M. (1984) 'Preparing nematode permanent mounts with adhesive tapes', *Journal of Nematology*, vol 16, no 3, pp341–342

Hunt, D. J. (1993) *Aphelenchida, Longidoridae, and Trichodoridae: Their Systematics and Bionomics*, CAB International, Wallingford

Jairajpuri, M. S. and Ahmad, W. (1992) *Dorylaimida: Free-living, Predaceous and Plant Parasitic Nematodes,* E. J. Brill, Leiden

Jenkins, W. R. (1964) 'A rapid centrifugal flotation technique for separating nematodes from soil', *Plant Disease Reporter*, vol 48, no 9, pp662–665

Krebs, C. J. (1994) *Ecology: The Experimental Analysis of Distribution and Abundance*, 4th edition, Harper Collins College Publishers, New York

Loof, P. A. (1991) 'The family Pratylenchidae Thorne, 1949', in W. R. Nickle (ed) *Manual of Agricultural Nematology*, Marcel Dekker, New York, pp363–422

Maggenti, A. R. (1991) 'Nemata: Higher classification', in W. R. Nickle (ed) *Manual of Agricultural Nematology*, Marcell Dekker, New York, pp147–190

Magurran, A. E. (1988) *Ecological Diversity and its Measurement*, Cambridge University Press, Cambridge

May, W. F., Mullin, P. G., Lyon, H. H. and Loeffer, K. (1996) *Pictorial Key to Genera of Plant Parasitic Nematodes*, 5th edition, Cornell University Press, New York

Nickle, W. R. and Hooper, D. J. (1991) 'The Aphelenchina: Bud, leaf, and insect nematodes', in W. R. Nickle (ed) *Manual of Agricultural Nematology*, Marcel Dekker, New York, pp465–508

Norton, D. C. (1978) *Ecology of Plant Parasitic Nematodes*, John Wiley and Sons, New York

Petersen, H. (1982) 'Structure and size of soil animal populations', *Oikos*, vol 39, no 3, pp306–329

Procter, D. L. C. (1990) 'Global overview of the functional roles of soil-living nematodes in terrestrial communities and ecosystems', *Journal of Nematology*, vol 22, no 1, pp1–7

Raski, D. J. (1991) 'Tylenchulidae in agricultural soils', in W. R. Nickle (ed) *Manual of Agricultural Nematology*, Marcel Dekker, New York, pp761–794

Seinhorst, J. W. (1959) 'A rapid method for the transfer of nematodes from fixative to anhydrous glycerin', *Nematologica*, vol 4, pp67–69

Smart, G. C., Jr and Nguyen, K. B. (1988) *Illustrated Key for Identification of Common Nematodes in Florida*, University of Florida, Gainesville

Smart, G. C., Jr and Nguyen, K. B. (1991) 'Sting and awl nematodes: *Belonolaimus* spp. and *Dolichodorus* spp.', in W. R. Nickle (ed) *Manual of Agricultural Nematology*, Marcell Dekker, New York, pp627–668

Sohlenius, B. (1980) 'Abundance, biomass and contribution to energy flow by soil nematodes in terrestrial ecosystems', *Oikos*, vol 34, no 2, pp186–194

Thorne, G. (1961) *Principles of Nematology*, McGraw Hill Co, New York

Yeates, G. W. (1994) 'Modification and quantification of the nematode maturity index', *Pedobiologia*, vol 38, pp97–101

Yeates, G. W., Bongers, T., de Goede, R. G. M., Freckman, D. W. and Georgieva, S. S. (1993) 'Feeding habits in soil nematode families and genera – an outline for soil ecologists', *Journal of Nematology*, vol 25, pp315–331

Chapter 6

Nitrogen-fixing Leguminosae-nodulating Bacteria

Fátima M. S. Moreira

Economic and ecological importance of nitrogen-fixing Leguminosae-nodulating bacteria (NFLNB) symbiosis

Biological nitrogen fixation is one of the most important processes for the maintenance of life on earth as it contributes about 70 per cent of all nitrogen required by natural and agricultural ecosystems (Burns and Hardy, 1975), and is environmentally friendly. Inoculation with NFLNB strains that are highly efficient and adapted to prevailing environmental conditions, to replace chemical N fertilizers, is currently practised in some countries for a small selection of legume crop species. In Brazil, inoculation with *Bradyrhizobium* strains completely replaces application of chemical fertilizers for soybean. In 2006, with a soybean yield of 57 million megagrams (Mg), about US$3.3 billion in fertilizer expenses were saved because of this biotechnology. Inoculant strains are sourced from a diversity of strains present in the soil. At the same time soil biodiversity may affect the nodulation behaviour positively or negatively, due to the complex and multiple interactions among soil organisms, and between soil organisms and plants. Thus, the knowledge of NFLNB diversity in the soil environment is the first step in the management and conservation of this valuable genetic resource.

Current taxonomy of NFLNB

Nitrogen-fixing bacteria, formerly known collectively as rhizobia, form nodules on roots (and exceptionally on stems) of some species of Leguminosae and on roots of *Parasponia* spp. (Ulmaceae). The term 'rhizobia' is related to Rhizobiaceae Conn 1938, a family-level taxon of bacteria created expressly to accommodate organisms able to nodulate Leguminosae. A significant recent extension of knowledge regarding nitrogen-fixing Leguminosae-nodulating bacteria was the discovery that bacteria belonging to the β-Proteobacteria (genera *Burkholderia* and *Ralstonia/Cupriavidus*) and to other families in the α-Proteobacteria (e.g. *Methylobacterium* Methylobacteriaceae and *Devosia* Hyphomicrobiaceae) were also able to nodulate Leguminosae (Chen et al, 2001; Moulin et al, 2001; Sy et al, 2001; Rivas et al, 2002, 2003; Jourand et al, 2004) (Table 6.1). For this reason the name 'rhizobia' is no longer suitable to describe nodulating bacteria, although its use continues in some literature.

Leguminosae comprise about 20,000 species distributed in the sub-families Caesalpinoideae (2250 spp., mainly woody tropical plants), Mimosoideae (3270 spp., mainly woody tropical, sub-tropical and temperate) and Papilionoideae (13,800 spp., largely herbaceous) (Lewis et al, 2005). The extent of the symbiosis with NFLNB among Leguminosae is still the subject of active research. Until 1989, only 57 per cent of genera and 20 per cent of species had been examined for nodulation, and the percentages of species able to nodulate were 23, 90 and 97 among Caesalpinoideae, Mimosoideae and Papilionoideae respectively (Faria et al, 1989). Although extensive searches for new nodulating genera and species have been made, especially in Brazil (Faria et al, 1989; Moreira et al, 1992), it is estimated that today only 25 per cent of the extant species have been examined. It is thus possible that many new symbioses will be identified in natural ecosystems (e.g. tropical forests). Traditionally, the taxonomy of NFLNB was based on strains isolated from temperate crop plants. However, now that isolates from other species and regions are available, and new techniques of molecular genetics have been developed, the taxonomy has been revolutionized, with 51 species and ten genera added to the original four species and two genera listed in 1984 by Jordan (see Table 6.1), and further revisions are expected.

EVALUATION OF NFLNB DIVERSITY IN SOIL

Studies of NFLNB populations and NFLNB diversity depend on successful isolation from root (or occasionally stem) nodules of legume host plants. Nodules may be sampled from field grown plants (in situ collections), although this approach may be difficult with woody or perennial hosts (Moreira et al, 1992; Odee et al, 1995). In order to isolate and enumerate NFLNB from a diverse microbial population, such as that which occurs in the soil, a method is required that clearly separates NFLNB from other bacterial species and allows easy and systematic sampling of nodules from nodulated plants. The plant infection technique makes use of the nodulation process itself to estimate NFLNB populations in soil. Further culture and characterization of the NFLNB from the nodules thus formed can provide information on the taxonomic composition of NFLNB populations and degrees of specificity between particular strains and candidate hosts. Although some host plants are considered highly promiscuous (i.e. show low NFLNB specificity), no one promiscuous host can be nodulated by all existing NFLNB species/strains, and conversely, there is no existing NFLNB strain sufficiently promiscuous to nodulate all legume species. Thus, to evaluate NFLNB diversity in soil it is desirable to make use of a variety of candidate host plant species, and the more that are employed, the greater will be the variety of NFLNB strains isolated and identified. The bioassay for NFLNB can make use of promiscuous hosts grown in soil sampled from the field or inoculated with soil suspensions (Moreira et al, 1993; Odee et al, 1997; Pereira,

Table 6.1 *Phylum/order, family, genera and species of Leguminosae-nodulating bacteria*

Phylum/Family/Genera/Species	*Reference*
α-Proteobacteria – Order Rhizobiales	
Rhizobiaceae	
Rhizobium	Frank, 1889
R. leguminosarum	Frank, 1879, 1889
Biovars phaseoli, trifolii, viceae	Jordan, 1984
R. galegae	Lindström, 1989
R. tropici	Martínez-Romero et al, 1991
R. etli	Segovia et al, 1993
R. giardinii biovars phaseoli, giardinii	Amarger et al, 1997
R. gallicum	Amarger et al, 1997
R. hainanense	Chen et al, 1997
R. mongolense	van Berkum et al, 1998
R. huautlense	Wang et al, 1998
R. etli biovar mimosae	Wang et al, 1999a
R. yanglingense	Tan et al, 2001
R. sullae	Squartini et al, 2002
R. indigoferae	Wei et al, 2002
R. loessense	Wei et al, 2003
R. daejeonense	Quan et al, 2005
Sinorhizobium	Chen et al, 1988; de Lajudie et al, 1994
S. meliloti	Dangeard, 1926; Casida, 1982; Jordan, 1984; de Lajudie et al, 1994; Young, 2003
S. fredii	Scholla and Elkan, 1984; Chen et al, 1988; de Lajudie et al, 1994
S. xinjiangense	Chen et al, 1988
S. saheli	de Lajudie et al, 1994
S. teranga	de Lajudie et al, 1994
S. medicae	Rome et al, 1996
S. arboris	Nick et al, 1999
S. kostiense	Nick et al, 1999
S. (Ensifer?) adhaerens	Casida, 1982; Willems et al, 2003; Young, 2003
S. morelense	Wang et al, 2002
S. kummerowiae	Wei et al, 2002
S. americanum	Toledo et al, 2003
Allorhizobium[1]	de Lajudie et al, 1998a
A. (Rhizobium) undicola	de Lajudie et al, 1998a; Young et al, 2001
Bradyrhizobiaceae	
Bradyrhizobium	Jordan, 1984
B. japonicum	Jordan, 1984
B. elkanii	Kuykendall et al, 1992
B. liaoningense	Xu et al, 1995
B. yuanmingense	Yao et al, 2002
B.canariense	Vinuesa et al, 2005a

Phylum/Family/Genera/Species	Reference
Blastobacter (Bradyrhizobium)	
B. denitrificans[2]	van Berkum and Eardly, 2002; van Berkum et al, 2006[2]
Xanthobacteraceae	
Azorhizobium	Dreyfus et al, 1988
A. caulinodans	Dreyfus et al, 1988
A. doebereinerae	Moreira et al, 2006
Phyllobacteriaceae	
Mesorhizobium	Jarvis et al, 1997
M. loti	Jarvis et al, 1982; Jordan, 1984; Jarvis et al, 1997
M. huakuii	Chen et al, 1991;Jarvis et al,1997
M. ciceri	Nour et al, 1994; Jarvis et al, 1997
M. tianshanense	Chen et al, 1995; Jarvis et al, 1997
M. mediterraneum	Nour et al, 1995; Jarvis et al, 1997
M. plurifarium	de Lajudie et al, 1998b
M. amorphae	Wang et al, 1999b
M. chacoense	Velásquez et al, 2001
M. temperatum	Gao et al, 2004
M. spetentrionale	Gao et al, 2004
Phyllobacterium	
Phyllobacterium trifolii	Valverde et al, 2005
Methylobacteriaceae	
Methylobacterium	
M. nodulans	Sy et al, 2001, Jourand et al, 2004
Brucellaceae	
Ochrobactrum	Trujillo et al, 2005
Ochrobactrum lupini	
β-Proteobacteria Order **Burkholderiales**	
Burkolderiaceae	
Burkholderia[3]	Moulin et al, 2001
B. tuberum	Vandamme et al, 2002
B. phymatum	Vandamme et al, 2002
B. caribensis	Vandamme et al, 2002
B. cepacia genomovar VI	Vandamme et al, 2002
B. mimosarum	Chen et al, 2006
Ralstonia (Cupriavidus)	
R.(Cupriavidus) taiwanensis[4]	Chen et al, 2001, Vandamme and Conye, 2004
Hyphomicrobiaceae	
Devosia	
D. neptunea	Rivas et al, 2002, 2003

Note: The references listed are those that describe each species.

1 *Agrobacterium* species *A. tumefasciens* (syn. *A. radiobacter*), *A. rhizogenes*, *A. rubi* and *A. vitis* and *Allorhizobium undicola* were proposed to be included in *Rhizobium* (Young et al, 2001).

2 These authors did not describe this species but they discovered it is able to nodulate legumes and they proposed to include the species in the genus *Bradyrhizobium*.

3 These authors did not describe the genus but discovered it is able to nodulate.

4 Genus *Wautersia* (Vaneechoutte et al, 2004) and later genus *Cupriavidus* (Vandamme and Conye, 2004) were proposed to accommodate this species.

2000) and the species found should then be compared with species nodulating naturally at the site. Although some of the latter associations may be relatively specific, the comparison of NFLNB isolated from naturally formed nodules with those sampled via the bioassay provides a useful check on the accuracy of the laboratory procedure. For instance, *Macroptilium atropurpureum* is one of the widely accepted promiscuous hosts (Vincent, 1970), but in most cases it is reported as being predominantly nodulated by *Bradyrhizobium* species (Woomer et al, 1988a), although it can also be nodulated by fast growers such as *Rhizobium* spp. (Pereira, 2000; Lima, 2007) and *Burkholderia* spp., among others (Moreira et al, 2002; Moreira, 2006, 2008). Preliminary results from Taita, a CSM-BGBD benchmark area in Kenya, indicated that *M. atropurpureum* was predominantly nodulated by slow-growing *Bradyrhizobium* spp. Similarly, Lewin et al (1987) demonstrated that *Vigna unguiculata*, usually considered a *Bradyrhizobium* host, in fact has very low specificity and can be nodulated by fast-growing *Rhizobium* spp. Results from Brazil (soil from the Alto Solimões river benchmark area) show that cowpea traps more species than *M. atropurpureum* and beans (*Phaseolus vulgaris*), but only siratro trapped NFLNB from all the sampling points. Preliminary experiments are therefore recommended for each soil environment before the final choice of host plants is made for the bioassay; where possible a native legume species should be included as trap host.

Nitrogenase is the enzyme responsible for the reduction of nitrogen gas to ammonia. It additionally reduces, among other substrates, acetylene to ethylene (Dilworth, 1966; Schollhorn and Burris, 1966). This reaction is used as a technique for the measurement of nitrogenase activity. The great advantage of acetylene reduction assay (ARA) is its great sensitivity and speed. It is also not expensive and relatively simple to carry out, even under field conditions. Although the use of ARA for quantitative estimates of N_2 fixation contribution to plant nutrition has been widely criticized (Boddey, 1987; Giller, 1987), it is very useful for the simple detection of N_2-fixers. For instance, as nodule anatomy varies widely in shape and size, the less experienced worker may confuse them with structures not induced by NFLNB. It can also be used for confirmation of new NFLNB symbiosis or for unusual types of nodule (Moreira et al, 1992). Nitrogenase activity constitutes valuable information because in many cases it is impossible to verify if nodules are still viable and effective (red colour inside) as nodules must remain intact for later isolation or must be dried if they have to be stored for longer time. Nodules without nitrogenase activity can be senescent or ineffective.

Methodology applied in the CSM-BGBD Project: An overview

Steps in the methodology for NFLNB characterizations applied in the CSM-BGBD project are set out below and summarized in Figure 6.1 as six steps and illustrated in Plates 3a to 3f. A previous description was given by Moreira (2004).

Step 1A: Collect soil samples from the field as described in the note.

Step 2A: In the project, the conventional method of using promiscuous trap species to isolate NFLNB from soil samples is employed. However, the use of multiple trap species

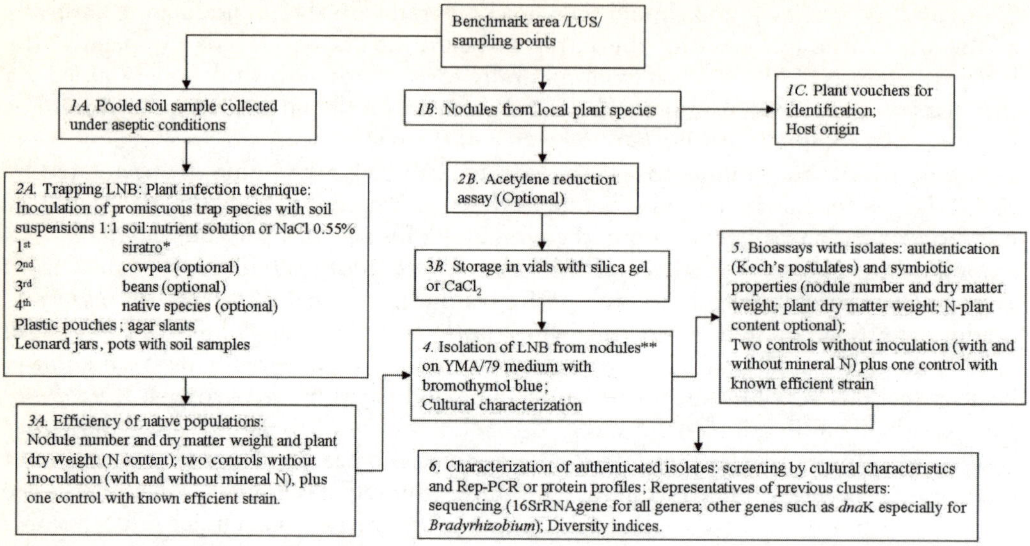

* Can be used for counting-MPN (optional);
** Some nodules can be kept on silica gel for further isolation if necessary.

Note: 1A. 12 cores in each sampling point (see overall CSM-BDBG scheme, Chapter 2). 2A. Trapping of LNB using the plant infection technique. 3A. Efficiency of native populations is assessed at the time of flowering or when significant differences can be observed among treatments. 4. Cultural characterization (preliminary experiment with forest and legume crop soil are needed to determine the number of isolates required to characterize the symbiont communities as a whole, based on species extinction curve). 6. Includes cultural characteristics obtained in step 4 for screening the whole collection of isolates (i.e. to get clusters) and Repetitive Extragenic Palindromic elements/DNA sequences (Rep-PCR) (or other method such as protein profiles by SDS-PAGE/polyacrylamide gel electrophoresis) can be complementary to this. Representatives of clusters, based on cultural characteristics and/or other method, must be sequenced for 16SrRNA and other genes such as dnaK (e.g. for *Bradyrhizobium* but can be applied for other genera).

Figure 6.1 *Evaluation of nitrogen-fixing Leguminosae-nodulating bacteria in diverse land use systems, summarized in six steps*

is not feasible with a large number of sampling points (e.g. 100) due to the laborious nature of the procedure and the limits of time and laboratory facilities. Therefore, a single promiscuous trap species is employed. In these circumstances, *Macroptilium atropurpureum* (siratro) is a good choice, as it has small seeds and is easy to manipulate under controlled conditions in plastic pouches and Leonard jars (with nutrient solution) in growth chambers or the greenhouse. When multiple experiments are carried out, seeds from the same accession should be used. This would also avoid any possible influence of plant accession on NFLNB trapped.

Where a team can handle more than one trap species, the additional promiscuous trap species recommended are *Vigna unguiculata* (cowpea, as the second species) and *Phaseolus vulgaris* (beans, as the third species). Seeds of locally available varieties/accessions will provide useful information for local farmers. The two species were chosen because they are quite promiscuous and also relevant as food crops in many countries. *Phaseolus vulgaris* is also selected because there are already some scattered data in several countries on NFLNB populations assessed with common bean as trap species, which could then act as a useful point of reference. After these three species (siratro, cowpea and beans), other species can be selected, including native species, which might of course be different between countries. Selection should be based on the criterion that the trap species is ecologically and economically relevant to the country. It would be impossible to standardize these choices.

Soil samples in pots can be used only if laboratories are available at the benchmark area, taking care to follow all normal microbiological aseptic techniques. Soils with high N content should be avoided; other limiting factors such as macro- and micronutrient deficiencies, acidity and high Al contents must be amended for assays. Trap species can be used to trap and count NFLNB, or only to trap them. As the first exercise is rather laborious, it is not mandatory, nor part of the standard methodology. To trap NFLNB, trap species can be cultivated in plastic pouches, Leonard jars or pots with soil samples as described below.

Step 3A: Although the efficiency under controlled conditions does not reflect what is happening in situ, because of different symbiotic relationships (different plant species) and environmental conditions, it is an easy measurement to make and gives useful information about the variability of the efficiency of the native populations. Besides, it can indicate the existence of efficient strains among soil populations for that particular host species.

Step 1B/2B, 3B: It is also recommended that NFLNB should be isolated from legume species (native or introduced) that nodulate naturally in the various land use systems. Comparison of NFLNB isolated from field nodules and from nodules induced on trap species by inoculating with soil suspensions will enable both a better evaluation of diversity and better assessment of the efficiency of the trap species. It is known that genetic diversity in natural populations of NFLNB from the same site may differ among isolates from nodules made in the field and isolates from nodules taken from plants grown in a soil dilution series in the laboratory (Bala et al, 2001).

Step 1C: Legume species in a given area (e.g. 8m radius) around each sampling point must therefore be inventoried in order that relationships with natural NFLNB populations can be ascertained, as explained before.

Step 4: After appearance and growth of nodules, NFLNB must be isolated from them. The isolations from these nodules and those from nodules collected from plants nodulating naturally in the various land use systems under consideration will constitute the diversity data. When the plant infection technique is used, nodules should be sampled from nodulated plants representing the sequential dilution levels (Moreira and Pereira, 2001), as NFLNB types trapped may also vary with dilution (Bala et al, 2001). Discussion in the literature seems to indicate that at least 30–50 nodules are needed overall, and possibly more than 100 (Jesus et al, 2005), permitting the construction of collection curves to assess whether diversity within a location is fully characterized. Considering that these curves are obtained based on the number of different cultural types that may still show variable genetic characteristics, a larger rather than a smaller number of isolates is necessary for good resolution. The collection (accumulation or rarefaction) curves can reveal the number of isolates needed to assess diversity properly.

This number is found where the curve reached the asymptote. If a plateau (asymptote) is not reached it indicates that not all species have been detected and that additional samples (nodules isolates) should be analysed. The collection curves must be applied to all isolates, that is, those from the bioassay and those from nodules collected in the field. Nodules are kept in silica gel (Figure 6.2) to make them available for possible additional isolation.

Isolation of NFLNB from nodules must be performed on medium 79 with bromothymol blue (Fred and Waksman, 1928). Medium 79 is also called YMA (Vincent, 1970) (Appendix 6.1). This enables a full cultural characterization (growth rate, pH change, exopolysaccharide production, colony morphology, colour and size, etc.). Clusters can be obtained by screening for these cultural characteristics. Genetic diversity of NFLNB populations can be also assessed by clustering of profiles that result from Rep-PCR, using software like 'Gelcompar' (de Bruijn et al, 1997) or by use of other techniques (e.g. protein profiles by SDS-polyacrylamide gel electrophoresis) that provide good resolution at strain level (**step 6**). However, this can be optional depending on the resources available in each country and must be carried out after **step 5** in order to avoid wasting time with contaminants. Finally, representatives of cultural clusters, Rep-PCR clusters or clusters obtained by other methods will be sequenced for 16SrRNA gene (**step 6**). If resources are available dnaK or other housekeeping genes such as atpD, rpoB, recA, glnII (Parker, 2004; Vinuesa et al, 2005b; Gaunt et al, 2001) can also be sequenced for strains of some genera which are not well discriminated by 16SrRNA (e.g. *Bradyrhizobium*).

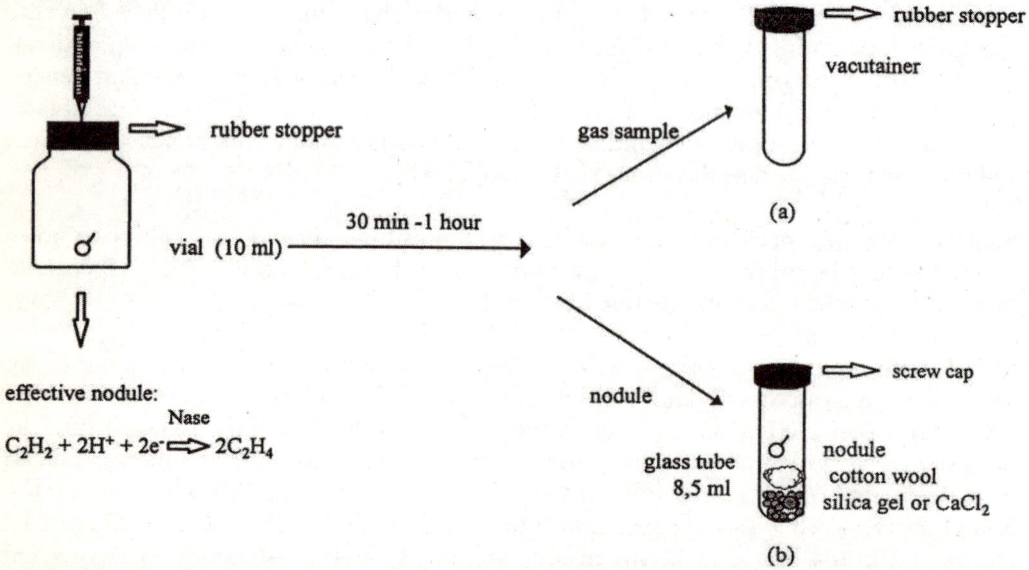

Source: From Moreira and Pereira, 2001

Figure 6.2 *Fieldwork: a) taking gas samples from nitrogenase-mediated acetylene reduction; and b) storage of nodules until isolation in the lab*

Material requirements for field and lab work

Fieldwork: for *soil sampling*: alcohol (95 per cent), water for washing off soil from sampling implements before alcohol sterilization, insulated cold box, sterilized plastic bags (300mL), spatula, large plastic bags (5L) and small soil corer; for *nodule sampling*: small scissors, spade, hoe, mattock, forceps, shovel, screw-cap tubes with silica gel or anhydrous $CaCl_2$; for *plant vouchers*: alcohol, press, old newspapers.

Laboratory work for NFLNB trapping and enumeration: 1mL and 5mL pipettes or micropipettes, diluent solution, 1L and 125mL Erlenmeyer flasks, orbital shaker, sterile plastic bags (125mL, as growth pouches), glass tubes (150×20mL or 200×30mm), or recycled beer bottles, racks for growth pouches or tubes, nutrient solution, seeds of promiscuous and/or native legume host plants, controlled environment room (temperature, light, humidity).

Laboratory work for NFLNB isolation and culture characterization: Petri dishes, 95 per cent alcohol, 0.1 per cent $HgCl_2$ (acidified with conc. HCl at 5mL/L) (Na or Ca hypochlorite, or H_2O_2 can be used to substitute $HgCl_2$), sterilized water, forceps, yeast–mannitol–mineral salts agar medium, pH 6.8.

Laboratory work for measuring nitrogenase activity in nodules (by acetylene reduction): Kitasato Erlenmeyer flasks, rubber ball (of the type used inside footballs), 1mL gas-tight syringes, 5mL vacutainers, 10mL (or larger) vials with rubber stoppers, calcium carbide (CaC_2), gas chromatograph equipped with flame ionization detector (FID) and Poropak RN column for acetylene/ethylene determinations. NB: nitrogenase assays can be performed in the field.

Methods in detail

Soil sampling: soil is cored to a depth of 20cm at 12 points distributed around each sampling plot as described in Chapter 2, Figure 2.3. The same coring scheme is used for sampling of all microbial groups, including NFLNB. Each set of 12 samples is bulked to form a composite sample of about 300g and placed inside a sterile plastic bag. Alternatively, if resources permit, three or more composite samples can be collected per sampling point. All sampling tools (corers, spatula, hoe, etc.) must be thoroughly washed with water to remove soil particles and flamed with alcohol before and after sampling at each sampling plot to avoid the introduction of exotic NFLNB. Trampling should be minimized close to the coring points and litter must be removed just before sampling takes place. Soil samples should then be transferred to the laboratory in an insulated container (preferably at 4°C) as soon as possible. A second bulk sample of about 200g should be collected in a non-sterile plastic bag for soil physical and chemical analysis.

Nodule sampling: leguminous plants within a circle of 8m radius (the same as for the vegetation inventory) from the sampling point should be identified and botanic material collected (see below). Prior information on which species are able to nodulate will be helpful, as the collection can then be confined to these species. However, it must be noted that there is a huge potential for the discovery of new nodulating legume species. For herbaceous plants, the whole root system can be removed from the soil (using hoe, spade or mattock as required), with care not to accidentally sever existing nodules. Nodules of woody plants must be discovered by excavation of the roots, taking care to explore the finer ramifications where nodulation is more commonly found. Extreme care must be taken to assure that fine roots belong to the individual plant identified as a legume. Thus, it is recommended that excavation starts by the trunk. The nodules are then excised (leaving a piece of root to facilitate manipulation) and stored individually in screw-cap tubes containing desiccant (Figure 6.2b). Before storage soil particles must be removed either by shaking or preferably by washing and removing water excess with a tissue. At least 50 nodules should be collected per sampling point, and the sample should be representative of all nodulating species occurring within the sampling plot. Occasionally, nodules may be too large for the ordinary screw-cap tubes and should be stored in a larger container.

Nitrogenase activity can be measured in the field on individual nodules, just after sampling, or in the laboratory (Figure 6.2). The nodule is put in a 10mL (or larger if needed) vial with a rubber stopper. Acetylene is produced in a Kitasato Erlenmeyer by the reaction of CaC_2 with water (Figure 6.3). If the work is carried out in the laboratory acetylene is obtained from commercially available gas cylinders. 1mL of this gas is injected into the nodule-containing vial. After one hour (or less) 1mL of headspace gas is removed and transferred to a vacutainer for the analysis of ethylene in the laboratory by gas chromatography (Figure 6.2a).

Plant voucher specimens: voucher specimens of nodulating plant species must be collected, in an area around the sampling point (8m radius), with careful attention to labelling (see below) and, if possible, the inclusion of flowers and fruits. The specimens should then be sent to a herbarium for identification, accompanied by an identification card as in the following example:

- **Project:** CSM-BGBD
- **Collector:** Fátima Moreira

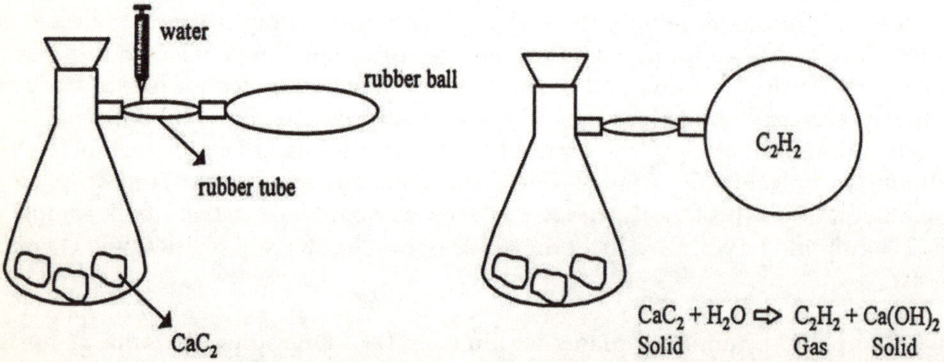

Figure 6.3 *Acetylene production in the field or laboratory*

- **Date:** 2 April 2005
- **Locality:** Benjamin Constant
- **Altitude:** 500m
- **Species vulgar name:** faveira
- **Species scientific name:**
- **Voucher number:** 05
- **Nodule characteristics:** indeterminate growth, size 0.5 to 1.5cm
- **Site description:** pasture, cattle grazing, open with few stumps
- **Soil type:** sandy loam
- **Plant characteristics:** herbaceous with mature fruits
- **Other comments:** seeds collected, yellow flowers

NFLNB counting: soil samples are submitted to serial dilutions before candidate host plants are inoculated (Figure 6.4). The dilution ratios vary between 2.0 and 14.5 depending on the expected concentration of cells in the soil sample (i.e. greater dilution for soils with more NFLNB). However, it is still necessary to inoculate host plants at all dilutions (see below). Also, at each dilution used, replication of the bioassay (2–5 times) should be employed. Plants are grown under controlled conditions (Vincent, 1970), and examined for nodule formation after 7–15 days. Populations of NFLNB are estimated by the most probable number method (Woomer et al, 1988b, 1990). For inoculating plants, when only trapping is intended, soil samples must be re-suspended in sterile water or nutrient solution (the same as used for culture medium or plant growth, see Appendices 6.1 and 6.2) at the ratio 1:1. Plants can be cultivated in plastic pouches, Leonard jars or other types of pot. Three types of controls without soil inoculum must be used. The first control without soil inoculum will check for the presence of contamination in the experimental procedures: if axenic conditions were not properly maintained this control will present nodules and the experiment is invalided (even if nodules occur in only one replication). It is recommended that the number of replications of this control should be greater than the replications of inoculation treatments, and the control should also be located in different positions in the protocol. The other two controls will allow to check the efficiency of LNB communities nodulating the host plant as well as whether the experimental conditions (e.g. temperature, nutrient concentrations) are suitable for nodulation and nitrogen fixation by plants and growth: the first of these controls is supplemented with mineral nitrogen but is not inoculated (for Leonard jars: 70mg $N\text{-}NH_4NO_3$ weekly, from the third up to the last-but-one week; for plastic pouches or small volume recipients: only one application of 70mg $N\text{-}NH_4NO_3$). Thus, for instance, cowpea, which has a growth period of two months in Leonard jars (up to flowering), will receive a total of 350mg $N\text{-}NH_4NO_3$, and beans, with a growth period of one and a half months (in Leonard jars), will receive 280mg $N\text{-}NH_4NO_3$. The other control will be inoculated with the recommended inoculant for the plant species (e.g. CIAT 899 for *Phaseolus vulgaris* in Brazil), but no nitrogen is added.

NFLNB isolation and characterization: NFLNB are isolated from nodules collected in the field and from those obtained under laboratory bioassay. In the latter case, nodules obtained at each successive dilution can indicate those strains that are rare in the soil sample, as well as the most common ones. The first step is to surface sterilize the nodules by a brief immersion in 95 per cent alcohol, followed a longer immersion up to 3–4

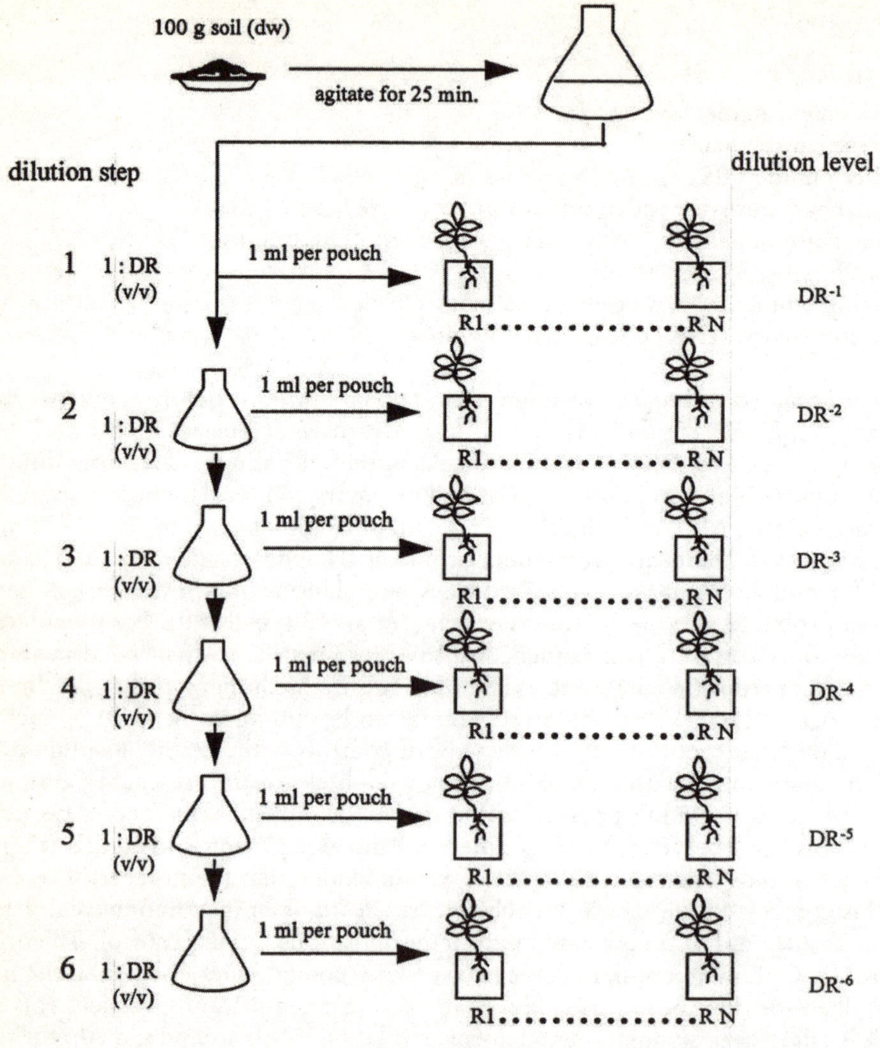

Note: Base dilution rates (DR) can vary from 2 to 14.5 and replicate numbers per dilution (N) from 2 to 5.
Source: Adapted from Woomer, 1993 and Vincent, 1970 by Moreira and Pereira, 2001

Figure 6.4 *Method for calculation of most probable number of rhizobia cells in soil by the plant infection technique*

minutes in $HgCl_2$ (Na or Ca hypochlorite, or H_2O_2 can be used to substitute) and washing in several rinses of sterile water (Vincent, 1970). The nodule is then crushed in a few drops of sterile water, using forceps, and a loopful of this suspension is streaked onto an agar medium. In the case of desiccated nodules, they should first be soaked in sterile water to improve their wettability by the sterilants. Immersion times in $HgCl_2$ (or in other disinfectants) should be adjusted to nodule size (shorter for smaller nodules). Composition of the yeast–mannitol–mineral salts agar medium (Fred and Waksman, 1928; Appendix 6.1), especially pH and carbohydrate source, can be varied to take account of particular soil conditions (Date and Halliday, 1979, 1987; Souza et al, 1984; Elkan and Bunn, 1991). Bromothymol blue can be included as an indicator, as pH changes caused by NFLNB growth may be useful in genus identification. Other characters include growth rate (TAIC-time of appearance of isolated colonies), the extent of extracellular polysaccharide deposition, colony shape, size, diameter and colour as described in Moreira et al (1993) and Jesus et al (2005).

The main generic descriptors are:

Allorhizobium, *Rhizobium* and *Sinorhizobium*: colonies circular, 2–4mm in diameter, but usually coalesce due to copious extracellular polysaccharide production, convex, semi-translucent, raised and mucilaginous, usually produce acid reaction, some with a yellowish centre (due to pH indicator), fast growers (TAIC: 2–3 days). *Mesorhizobium*: same as *Rhizobium*, but usually intermediate growers (TAIC: 4–5 days).

Bradyrhizobium: colonies circular, not exceeding 1mm in diameter, extracellular polysaccharide production from abundant to little (the latter generally in those strains taking more than ten days to grow), opaque, rarely translucent, white and convex, granular in texture, usually producing an alkaline pH shift, slow or very slow growers (TAIC: six or more days).

Azorhizobium: colonies circular, 0.5mm in diameter with a creamy colour, very little extracellular polysaccharide production (much less than in *Bradyrhizobium*), produce an alkaline pH shift, fast to intermediate growers (TAIC: 3–4 days).

Cupriavidus: similar to *Azorhizobium* with slightly more extracellular polysaccharide production.

Burkholderia: growth characteristics similar to fast growers (e.g. *Rhizobium*) except for pH modification as they can produce acid and alkaline reaction depending on age and sometimes at the same time.

NFLNB are non-spore-forming G-rods, usually containing poly-β-hydroxybutyrate granules refractile under phase contrast microscopy. Isolates must be reconfirmed as NFLNB, by a demonstration that they will again nodulate a test host plant under bacteriological controlled conditions, following Koch's postulates. Table 6.1 gives references that provide full details of NFLNB species characteristics. Strain diversity within species may be high, both genetically and phenotypically (for example symbiotic, cultural, morphological and physiological traits) and it is thus necessary to define the level of diversity which is appropriate to characterize particular genera and species. Minimum standards for the description of new genera and species were given by Graham et al (1991). These standards included: serology (Dudman and Belbin, 1988), cell lipopolysaccharides (de Maagd et al, 1988), total protein patterns by SDS-PAGE (Dreyfus et al, 1988; Moreira et al, 1993), small subunit RNA sequence (Young and Haukka, 1996), plasmid profiles (Giller et al, 1989), intrinsic antibiotic resistance (IAR) (Kingsley and Bohlool, 1983), multilocus enzyme electrophoresis (Selander et al, 1986), growth on different C sources (Dreyfus et al, 1988), DNA:DNA relatedness, DNA base composition (percentage C + G), rRNA:DNA hybridization and DNA restriction fragment length polymorphism (RFLP). More recently, new techniques for characterization have been described, improving not only bacterial classification but also the discrimination capability within different taxa: each of them has a specific level of resolution for bacterial classification, which might be useful for diversity studies. These techniques, also called fingerprint approaches, include: digestion of genomic DNA with rare cutting site endonucleases, followed by PFGE (pulse field gel electrophoresis) and other RFLP-based methods; polymerase chain reaction (PCR)-based methods, such as ARDRA (amplified rDNA restriction analysis), tRNA-PCR or ITS (amplification and analysis of inter tRNA spacer regions or inter 16S-23S rRNA gene regions), AFLP (amplified fragment length polymorphism) for the whole-genome analysis, RAPD (random amplified polymorphic DNA; AP-PCR (arbitrarily primed PCR), ReP-PCR (repetitive DNA elements genomic fingerprinting) (de Bruijn et al, 1997; Rademaker and de Bruijn, 1997). Numerical cluster analysis of an adequate number of strains and comparison with NFLNB type strains can permit the characterization of large populations. Genetic characterization (DNA, 16S rDNA, 23S rDNA or other gene sequencing; DNA:DNA homology) is time-consuming and requires specialized equipment, and is usually confined to representatives of clusters only. On the other hand, if resources are available, sequencing of specific genes can be applied directly to a number of strains determined by rarefaction, accumulation or collection curves. In this case, the costs and benefits of previous screening by other characteristics against further sequencing of representatives or the direct sequencing of a (probably) larger number of isolates must be taken into consideration. Costs of DNA sequencing have decreased a lot in recent years, thus in some cases direct sequencing of 16S rRNA or other genes would offer economic advantages. Private enterprises are offering good services for PCR products, its purification and sequencing. For instance, the purification and sequencing of one gene fragment (one sample) already amplified by PCR currently may cost about US$6 if complete plates (96 samples) are ordered. Total plates (96 samples) cost about US$300 for sequencing.

DNA extraction: genomic DNA is isolated from log phase cultures grown on 79/YMA for varying incubation periods depending on the specific growth rate of each strain (2–10 days). Ultra-clean Soil DNA isolation kits from MOBIO laboratories, or any other kit, can be used, essentially as recommended by the manufacturer. DNA is quantified at A_{260} nm with a spectrophotometer or estimated by comparison with different DNA concentration standards on agarose gels. Alternatively, for each strain a loopful of bacterial mass from an isolated colony is suspended in 1mL of sterile water to give a slightly turbid suspension and boiled for five minutes to lyse the cells. A volume of this suspension is used as a template for PCR.

16S rDNA sequencing: from isolates of representative clusters (obtained either by cultural characterization, Rep-PCR profiles or other techniques), near full-length 16S rDNA genes are amplified with primer pair 27F (pA:5'AGAGTTTGATCCTGGCTCAG) and 1492R (5'GGTTACCTTGTTACGACTT). These primer pairs correspond to positions 8–27 and to 1507–1492 respectively, of *Escherichia coli* 16S rDNA gene (Wilson et al, 1990). Other primers can be used since they target near full-length 16S rDNA gene. The final concentrations in the reaction mixtures (100mL) are: 1× PCR buffer, 2.5mM $MgCl_2$, 0.2mM of each dNTP, 0.2μM of each primer and two units of Taq polymerase template (DNA extracted or liquid cultures): 6μL. The PCR programme has an initial denaturing step at 94°C for five minutes; followed by 30 cycles of denaturing at 94°C for 40 seconds, annealing at 55°C for 40 seconds, and extension at 72°C for 90 seconds. The final extension is performed at 72°C for seven minutes. Purification of PCR products is performed with Microcon™ filters (Millipore) or other purification methods. Single pass sequencing of PCR-amplified rDNAs is performed with the 27F primer and also with the reverse primer 1492R. If sequencing the whole gene is required, internal primers should be chosen to amplify and sequence the gene fragment lacking after use of 27F and 1492R.

Phylogenetic analysis: the sequences are compared to known sequences contained in databases such as the National Center for Biotechnology Information (NCBI) (www.ncbi.nlm.nih.gov/), the European Bioinformatics Institute (EBI) sequences database (www.ebi.ac.uk) and the Ribosomal Database Project-RDP (http://rdp.cme.msu.edu/index.jsp).

MAINTENANCE AND COLLECTIONS OF PURE CULTURES

Usually pure cultures are maintained in agar slants with the same culture medium in screw-cap tubes, like those presented in Figure 6.2. However, viability of these cultures is not very long and methods such as lyophilization and storage in a deep freezer (–80°C) guarantee long-term storage of viable cells.

ACKNOWLEDGEMENT

We are grateful to Drs Esperanza Martínez, David Odee and Nancy Karanja for comments and/or revisions.

REFERENCES

Amarger, N., Macheret, V. and Laguerre, G. (1997) '*Rhizobium gallicum* sp. nov. and *Rhizobium giardinii* sp. nov., from *Phaseolus vulgaris* nodules', *International Journal of Systematic Bacteriology*, vol 47, pp996–1006

Bala, A., Murphy, P., Giller, K. E. (2001) 'Genetic diversity of rhizobia varies from natural populations with the soil dilution sampled', *Soil Biology and Biochemistry*, vol 33, pp841–843

Berkum, P. van and Eardly, B. D. (2002) 'The aquatic budding bacterium *Blastobacter denitrificans* is a nitrogen-fixing symbiont of *Aeschinomene indica*', *Applied and Environmental Microbiology*, vol 68, pp1132–1136

Berkum, P. van, Beyene, D., Bao, G., Campbell, T. A. and Eardly, B. (1998) '*Rhizobium mongolense* sp. nov. is one of three rhizobial genotypes identified which nodulate and form nitrogen fixing symbioses with *Medicago ruthenica* [(L.) Ledebour]', *International Journal of Systematic Bacteriology*, vol 48, pp13–22

Berkum, P. van, Leibold, J. M. and Eardly, B. D. (2006) 'Proposal for combining *Bradyrhizobium* spp. (*Aeschynomene indica*) with *Blastobacter denitrificans* and to transfer *Blastobacter denitrificans* (Hirsh and Muller, 1985) to the genus *Bradyrhizobium* as *Bradyrhizobium denitrificans* (comb. nov.)', *Systematic and Applied Microbiology*, vol 29, pp207–215

Boddey, R. M. (1987) 'Methods for quantification of nitrogen fixation associated with Graminae', *Critical Reviews in Plant Sciences*, vol 16, pp209–266

Bruijn, F. J. de, Davey, M. E., McSpadden-Gardener, B., Millcamps, A., Rademaker, L. W., Ragatz, D., Schultz, M. L., Struffi, P. and Stoltzfus, J. (1997) 'Molecular approaches in microbial ecology to assess genomic diversity and stress-induced gene expression in plant-associated diazotrophs', in C. Elmerich, A. Kondorosi and W. E. Newton (eds) *Biological Nitrogen Fixation for the 21st Century*, Kluwer Academic Publishers, Dordrecht

Burns, R. C. and Hardy, R. W. F. (1975) *Nitrogen Fixation in Bacteria and Higher Plants*, Springer-Verlag, New York

Casida, L. E., Jr (1982) '*Ensifer adhaerens* gen. nov., sp. nov.: A bacterial predator of bacteria in soil', *International Journal of Systematic Bacteriology*, vol 32, pp339–345

Chen, W. M., James, E. K., Coenye, T., Chou, J., Barrios, E., Faria, S. M., Elliot, G. N., Shen, S., Sprent, J. I .and Vandamme, P. (2006) '*Burkholderia mimosarum* sp. nov. isolated from nodules of *Mimosa* spp. from Taiwan and South America', *International Journal of Systematic and Evolutionary Microbiology*, vol 56, pp1847–1851

Chen, W. M., Laevens, S., Lee, T. M., Coenye, T., Vos, P., Mergeay, M. and Vandamme, P. (2001) '*Ralstonia taiwanensis* sp. nov. isolated from nodules of *Mimosa* species and sputum of a cystic fibrosis patient', *International Journal of Systematic and Evolutionary Microbiology*, vol 51, pp1729–1735

Chen, W. X., Li, G. S., Qi, Y. L., Wang, E. T., Yuan, H. L. and Li, J. L. (1991) '*Rhizobium huakuii* sp. nov. isolated from the root nodules of *Astragalus sinicus*', *International Journal of Systematic Bacteriology*, vol 41, pp275–280

Chen, W. X., Tan, Z. Y., Gao, J. L., Li, Y. and Wang, E. T. (1997) '*Rhizobium hainanense* sp. nov., isolated from tropical legumes', *International Journal of Systematic Bacteriology*, vol 47, pp870–873

Chen, W. X., Wang, E., Wang, S., Li, Y., Chen, X. and Li, Y. (1995) 'Characteristics of *Rhizobium tianshanense* sp. nov., a moderately and slowly growing root nodule bacterium isolated from an arid saline environment in Xinjiang, People's Republic of China', *International Journal of Systematic Bacteriology*, vol 45, pp153–159

Chen, W. X., Yan, G. H. and Li, J. L. (1988) 'Numerical taxonomic study of fast-growing soybean rhizobia and a proposal that *Rhizobium fredii* be assigned to *Sinorhizobium* gen. nov.', *International Journal of Systematic Bacteriology*, vol 38, pp392–397

Dangeard, P. A. (1926) 'Recherches sur les tubercles radicaux des légumineuses', *Botaniste (Paris)*, vol 16, pp1–275

Date, R. A. and Halliday, J. (1979) 'Selecting *Rhizobium* for acid, infertile soils of the tropics', *Nature,* vol 277, pp62–64

Date, R. A. and Halliday, J. (1987) 'Collection, isolation, cultivation and maintenance of rhizobia' , in G. H. Elkan (ed) *Symbiotic Nitrogen Fixation Technology*, Marcel Dekker Inc, New York

Dilworth, M. J. (1966) 'Acetylene reduction by nitrogen fixing preparations of *Clostridium pasteurianum*', *Biochemica et Bioophysica Acta*, vol 127, pp285–294

Dreyfus, B., Garcia, J. L. and Gillis, M. (1988) 'Characterization of *Azorhizobium caulinodans* gen. nov., sp. nov., a stem-nodulating nitrogen-fixing bacterium isolated from *Sesbania rostrata*', *International Journal of Systematic Bacteriology*, vol 38, pp89–98

Dudman, W. F. and Belbin, L. (1988) 'Numerical taxonomic analysis of some strains of *Rhizobium* spp. that uses a qualitative coding of immunodiffusion reactions', *Applied and Environmental Microbiology*, vol 54, pp1825–1830

Elkan, G. H. and Bunn, C. R. (1991) 'The Rhizobia', in A. Balows, M. Dworkin, H. G. Schlegel and H. Truper (eds), *The Prokaryotes*, 2nd edition, Springer-Verlag, Berlin

Faria, S. M., Lewis, G. P., Sprent, J. I. and Sutherland, J. M. (1989) 'Occurrence of nodulation in the Leguminosae', *New Phytologist*, vol 111, pp607–619

Frank, B. (1879) 'Über die Parasiten in den Wurzelan-schwillungen der Papilionaceen', *Botanik Zeitung*, vol 37, pp376–387, 394–399

Frank, B. (1889) 'Über die Pilzsymbiose der Leguminosen', *Berichte der Deutschen Botanischem Gesellschaft*, vol 7, pp332–346

Fred, E. B. and Waksman, S. A. (1928) *Laboratory Manual of General Microbiology*, McGraw-Hill Book Company, Inc, New York and London

Gao, J. L., Turner, S. L., Kan, F. L., Wang, E. T., Tan, Z. Y., Qiu, Y. H., Terefework, Z., Young, J. P. W., Lindstrom, K. and Chen, W. X. (2004) '*Mesorhizobium septentrionale* sp. nov. and *Mesorhizobium temperatum* sp. nov. isolated from *Astragalus adsurgens* growing in the northern regions of China', *International Journal of Systematic and Evolutionary Microbiology*, vol 5, pp2003–2012

Gaunt, M. W., Turner, S. L., Rigottier-Gois, L., Lloyd-Macgilp, S. A. and Young, J. P. W. (2001) 'Phylogenies of *atp*D and *rec*A support the small subunit rRNA-based classification of rhizobia', *International Journal of Systematic and Evolutionary Microbiology*, vol 51, pp2037–2048

Giller, K. E. (1987) 'Use and abuse of the acetylene reduction assay for measurement of "associative" nitrogen fixation', *Soil Biology and Biochemistry*, vol 19, pp783–784

Giller, K. E., McGrath, S. P. and Hirsch, P. R. (1989) 'Absence of nitrogen fixation in clover grown on soil subject to long-term contamination with heavy metals is due to survival of only ineffective *Rhizobium*', *Soil Biology and Biochemistry*, vol 21, pp 841–848

Graham, P. H., Sadowsky, M. J., Keyser, H. H., Barnet, Y. M., Bradley, R. S., Cooper, J. E., de Ley, J., Jarvis, B. D. W., Roslycky, E. B., Strijdom, B. W. and Young, J. P. W. (1991) 'Proposed minimum standards for the description of new genera and species of root- and stem-nodulating bacteria', *International Journal of Systematic Bacteriology*, vol 41, pp582–587

Jarvis, B. D. W., van Berkum, P., Chen, W. X., Nour, S. M., Fernandez, M. P., Cleyet-Marel, J. C. and Gillis, M. (1997) 'Transfer of *Rhizobium loti, Rhizobuim huakuii, Rhizobium ciceri, Rhizobium mediterraneum* and *Rhizobium tianshanense* to *Mesorhizobium* gen. nov.', *International Journal of Systematic Bacteriology*, vol 47, 895–898

Jarvis, B. D. W., Pankhurst, C. E. and Patel, J. J. (1982) '*Rhizobium loti*, a new species of legume root nodule bacteria', *International Journal of Systematic Bacteriology*, vol 32, pp378–380

Jesus, E. C., Florentino, L. A., Rodrigues, M. I. D., Oliveira, M. and Moreira, F. M. S. (2005) 'Leguminosae nodulating bacteria diversity from three different land use systems in Brazilian Western Amazon', *Pesquisa agropecuária Brasileira*, vol 40, pp769–776

Jordan, D. C. (1984) '*Rhizobiaceae* Conn 1938', in N. R. Krieg and J. D. Holt (eds) *Bergey's Manual of Systematic Bacteriology*, Williams and Wilkins, London

Jourand, P., Giraud, E., Bena, G., Sy, A., Willems, A., Gillis, M., Dreyfus, B. and de Lajudie, P. (2004) '*Methylobacterium nodulans* sp. nov., for a group of aerobic facultatively methylotrophic, legume root-nodule-forming and nitrogen fixing bacteria', *International Journal of Systematic and Evolutionary Microbiology*, vol 54, pp2269–2273

Kingsley, M. T. and Bohlool, B. B. (1983) 'Characterization of *Rhizobium* sp. (*Cicer arietinum* L.) by immunofluorescence, immunodiffusion and intrinsic antibiotic resistance', *Canadian Journal of Microbiology*, vol 29, pp518–526

Kuykendall, L. D., Saxena, B., Devine, T. E. and Udell, S. (1992) 'Genetic diversity in *Bradyrhizobium japonicum* and a proposal for *Bradyrhizobium elkanii* sp. nov.', *Canadian Journal of Microbiology*, vol 38, pp501–505

Lajudie, P. de, Laurent-Fulele, E., Willems, A., Torck, U., Coopman, R., Collins, M. D., Kersters, K., Dreyfus, B. and Gilles, M. (1998a) '*Allorhizobium undicola* gen. nov., sp. nov. nitrogen-fixing bacteria that efficiently nodulate *Neptunia natans* in Senegal', *International Journal of Systematic Bacteriology*, vol 48, pp1277–1290

Lajudie, P. de, Willems, A., Nick, G., Moreira, F., Molouba, F., Hoste, B., Torck, U., Neyra, M., Collins, M. D., Lindstrom, K., Dreyfus, B. and Gillis, M. (1998b) 'Characterization of tropical tree rhizobia and description of *Mesorhizobium plurifarium* sp. nov.', *International Journal of Systematic Bacteriology*, vol 48, pp369–382

Lajudie, P. de, Willems, A., Pot, B., Dewettinck, D., Maestrojuan, G., Neyra, M., Collins, M. D., Dreyfus, B., Kersters, K. and Gillis, M. (1994) 'Polyphasic taxonomy of rhizobia: Emendation of the genus *Sinorhizobium* and description of *Sinorhizobium meliloti* com. nov.; *Sinorhizobium saheli* sp. nov.; and *Sinorhizobium teranga* sp. nov.', *International Journal of Systematic Bacteriology*, vol 44, pp715–733

Lewin, A., Rosenberg, C., Meyer, H., Wong, C. H., Nelson, L., Manen, J. F., Stanley, J., Dowling, D. N., Dénarie, J. and Broughton, W. J. (1987) 'Multiple host-specificity loci of the broad host-range *Rhizobium* sp. NGR234 selected using the widely compatible legume *Vigna unguiculata*', *Plant Molecular Biology*, vol 8, pp447–459

Lewis, G., Schrire, B., Mackinder, B. and Lock, M. (2005) *Legumes of the World*, Royal Botanic Gardens, Kew, London

Lima, A. S. (2007) 'Densidade, eficiência e diversidade de bactérias fixadoras de N2 que nodulam o siratro (*Macroptilium atropurpureum*) de solos sob diferentes usos na Amazônia Ocidental', PhD Thesis, Universidade Federal de Lavras, Minas Gerais, Brasil

Lindström, K. (1989) '*Rhizobium galegae*, a new species of legume root nodule bacteria', *International Journal of Systematic Bacteriology*, vol 39, pp365–367

Maagd, R. A. de, van Rossum, C. and Lugtenberg, J. J. (1988) 'Recognition of individual strains of fast-growing rhizobia by using profiles of membrane proteins and lipopolysaccharides', *Journal of Bacteriology*, vol 170, pp3782–3785

Martínez-Romero, E., Segovia, L., Mercante, F. B., Franco, A. A., Graham, P. and Pardo, M. A. (1991) '*Rhizobium tropici*, a new species nodulating *Phaseolus vulgaris* L. beans and *Leucaena* trees', *International Journal of Systematic Bacteriology*, vol 41, pp417–426

Moreira, F. M. S. (2004) 'Methodology for the inventory of nitrogen-fixing Leguminosae nodulating bacteria', Conservation and Sustainable Management of Below-ground Biodiversity CSM-BGBD, Annual Meeting Report 4, Nairobi, Kenya

Moreira, F. M. S. (2006) 'Nitrogen-fixing Leguminosae nodulating bacteria', in F. M. S. Moreira, J. O. Siqueira and L. Brussaard (eds) *Soil Biodiversity in Amazonian and other Brazilian Ecosystems*, CAB International, Wallingford

Moreira, F. M. S. (2008) 'Bactérias fixadoras de nitrogênio que nodulam Leguminosae', in F. M. S. Moreira, J. O. Siqueira and L. Brussaard (eds) *Biodiversidade do solo em ecossistemas brasileiros*, Editora, UFLA, Lavras

Moreira, F. M. S. and Pereira, E. G. (2001) 'Microsymbionts: Rhizobia', in M. Swift and D. Bignell (eds) *Standard Methods for Assessment of Soil Biodiversity and Land Use Practice*, International Centre for Research in Agroforestry, Bogor, Indonesia (available at http://fao.org/AG/agl/agll/soilbiod/docs/manual-soil%2)

Moreira, F. M. S., Cruz, L. M., Faria, S. M., Marsh, T., Martínez-Romero, E., Pedrosa, F. O., Pitard, R. and Young, P. J. W. (2006) '*Azorhizobium doebereinerae* sp. nov. microsymbiont of *Sesbania virgata* (Caz.) Pers.', *Systematic and Applied Microbiology*, vol 29, pp197–206

Moreira, F. M. S., Gillis, M., Pot, B., Kersters, K. and Franco, A. A. (1993) 'Characterization of rhizobia isolated from different divergence groups of tropical Leguminosae by comparative polyacrylamide gel electrophoresis of their total proteins, *Systematic and Applied Microbiology*, vol 16, pp135–146

Moreira, F. M. S., Silva, M. F. and Faria, S. M. (1992) 'Occurrence of nodulation in legume species in the Amazon region of Brazil', *New Phytologist*, vol 121, pp563–570

Moreira, F. M. S., Tiedje, J. and Marsh, T. L. (2002) '*Burkholderia* spp. are among fast growing symbiotic diazotrophs isolated from diverse land use systems in Amazonia and from Brazilian Leguminosae forest species', Memorias da XXI Reunión Latinoamericana de rhizobiologia, Cocoyoc, Mexico

Moulin, L., Munive, A., Dreyfus, B., and Bolvin-Masson, C. (2001) 'Nodulation of legumes by members of the b sub-class of *Proteobacteri*', *Nature*, vol 411, pp948–950

Nick, G., de Lajudie, P., Eardly, B. D., Suomalainen, S., Paulin, L., Zhang, X., Gillis, M. and Lidstrom, K. (1999) '*Sinorhizobium arboris* sp. nov. and *Sinorhizobium kostiense* sp. nov., isolated from leguminous trees in Sudan and Kenya', *International Journal of Systematic Microbiology*, vol 49, pp1359–1368

Nour, S. M., Cleyet-Marel, J. C., Beck, D., Effosse, A. and Fernandes, M. P. (1994) 'Genotypic and phenotypic diversity of *Rhizobium* isolated from chickpea (*Cicer arietnum* L.)', *Canadian Journal of Microbiology*, vol 40, pp345–354

Nour, S. M., Cleyet-Marel, J. C., Normand, P. and Fernandes, M. P. (1995) 'Genomic heterogeneity of strains nodulating chickpea (*Cicer arietnum* L.) and description of *Rhizobium mediterraneum* sp. nov.', *International Journal of Systematic Bacteriology*, vol 45, pp640–648

Odee, D. W., Sutherland, J. M., Kimiti, J. M. and Sprent, J. I. (1995) 'Natural rhizobial populations and nodulation status of woody legumes growing in diverse Kenyan conditions', *Plant and Soil*, vol 173, pp211–224

Odee, D. W., Sutherland, J. M., Makatiani, E. T., McInroy, S. G. and Sprent, J. I. (1997) 'Phenotypic characteristics and composition of rhizobia associated with woody legumes growing in diverse Kenyan conditions', *Plant and Soil*, vol 188, pp65–75

Parker, M. A. (2004) 'rRNA and *dna*K relationships of *Bradyrhizobium* sp. nodule bacteria from four Papilionoid legume trees in Costa Rica', *Systematic and Applied Microbiology*, vol 27, pp334–342

Pereira, E. G. (2000) 'Diversity of rhizobia isolated from different land use systems in Amazon region', PhD Thesis, Universidade Federal de Lavras, Lavras, Brazil

Quan, Z., Bae, H., Baek, J., Chen, W., Im, W. and Lee, S. (2005) '*Rhizobium daejeonense* sp. nov. isolated from a cyanide treatment bioreactor', *International Journal of Systematic and Evolutionary Microbiology*, vol 55, pp2543–2549

Rademaker, J. L. W. and de Bruijn, F. D. (1997) 'Characterization and classification of microbes by rep-PCR genomic fingerprinting and computer-assisted pattern analysis', in G. Caetano-Anollés, and P. M. Gresshoff (eds) *DNA Markers: Protocols, Applications, and Overviews*, John Wiley and Sons Inc, New York

Rivas, R., Velázquez, E., Willems, A., Vizcaíno, N., Subba-Rao, N., Mateos, P. F., Gillis, M., Dazzo, F. D. and Martínez-Molina, E. (2002) 'A new species of *Devosia* that forms a unique nitrogen-fixing root-nodule symbiosis with the aquatic legume *Neptunia natans* (L.f.) Druce', *Applied and Environmental Microbiology*, vol 68, pp5217–5222

Rivas, R., Willems, A., Subba-Rao, N., Mateos, P. F., Dazzo, F. D., Kroppenstedt, R. M., Martínez-Molina, E., Gillis, M. and Vizcaíno, N. (2003) 'Description of *Devosia neptuniae* sp. nov. that nodulates and fixes nitrogen in symbiosis with *Neptunia natans*, an aquatic legume from India', *Systematic and Applied Microbiology*, vol 26, pp47–53

Rome, S., Fernadez, M. P., Brunel, B., Normand, P. and Cleyet-Marel, J. C. (1996) '*Sinorhizobium medicae* sp. nov., isolated from annual *Medicago* spp.', *International Journal of Systematic Bacteriology*, vol 46, pp972–980

Scholla, M. H. and Elkan, G. H. (1984) '*Rhizobium fredii* sp. nov., a fast-growing species that effectively nodulates soybeans', *International Journal of Systematic Bacteriology*, vol 34, pp484–486

Schollhorn, R. and Burris, R. H. (1966) 'Study of intermediates in nitrogen fixation', *Proceedings of the Federation of American Societies for Experimental Biology*, vol 24, p710

Segovia, L., Young, J. P. W. and Martínez-Romero, E. (1993) 'Reclassification of American *Rhizobium leguminosarum* biovar phaseoli type I strains as *Rhizobium etli* sp. nov.', *International Journal of Systematic Bacteriology*, vol 43, pp374–377

Selander, R. K., Caugant, D. A., Ochman, H., Musser, J. M., Gilmour, M. N. and Whittam, T. S. (1986) 'Methods of multilocus enzyme electrophoresis for bacterial population genetics and systematics', *Applied and Environmental Microbiology*, vol 51, pp873–884

Souza, L. A. G., Magalhaes, F. M. M. and Oliveira, L. A. (1984) 'Avaliaçao do crescimento de rhizobium de leguminosas florestais tropicais em diferentes meios do cultura', *Pesquisa*

Agropecuária Brasileira – Ediçao Especial 19, s/n
Squartini, A., Struffi, P., Döring, H., Selnska-Pobell, S., Tola, E., Giacomini, A., Vendramin, E., Velázquez, E., Mateos, P. F., Martínez-Molina, E., Dazzo, F. B., Casella, S. and Nuti, M. P. (2002) '*Rhizobium sullae* sp. nov. (formerly '*Rhizobium hedysari*'), the root-nodule microsymbiont of *Hedysarum coronarium* L.', *International Journal of Systematic and Evolutionary Microbiology*, vol 52, pp1267–1276
Sy, A., Giraud, E., Jourand, P., Garcia, N., Willems, A., de Lajudie, P., Prin, Y., Neyra, M., Gillis, M., Boivin-Masson, C. and Dreyfus, B. (2001) 'Methylotrophic *Methylobacterium* bacteria nodulate and fix nitrogen in symbiosis with legumes', *Journal of Bacteriology*, vol 183, pp214–220
Tan, Z. Y., Kan, G. X., Wang, E. T., Reinhold-Hurek, B. and Chen, W. X. (2001) '*Rhizobium yanglingense* sp. nov. isolated from arid and semi-arid regions in China', *International Journal of Systematic and Evolutionary Microbiology*, vol 51, pp901–914
Toledo, I., Lloret, L. and Martínez-Romero, E. (2003) '*Sinorhizobium americanum* sp. nov., a new *Sinorhizobium* species nodulating Acacia spp. in Mexico', *Systematic and Applied Microbiology*, vol 26, pp54–64
Trujíllo, M. E., Willens, A., Abril, A., Planchuelo, A., Rivas, R., Ludeña, D., Mateos, P. F. and Martínez-Molina, E. (2005) 'Nodulation of *Lupinus albus* by strains of *Ochrobactrum lupini* sp. nov.', *Applied and Environmental Microbiology*, vol 71, pp1318–1327
Valverde, A., Velázquez, E., Fernández-Santos, F., Vizcaíno, N., Rivas, R. and Mateos, P. F. (2005) '*Phylobacterium trifolii* sp. nov. nodulating *Trifolium* and *Lupinus* in Spanish soils', *International Journal of Systematic Bacteriology*, vol 55, pp1985–1989
Vandamme, P. and Conye, T. (2004) 'Taxonomy of the genus *Cupriavidus*: A tale of lost and found', *International Journal of Systematic and Evolutionary Microbiology*, vol 54, pp2285–2289
Vandamme, P., Goris, J., Chen, W.-M., de Vos, P. and Willems, A. (2002) '*Burkholderia tuberum* sp. nov. and *Burkholderia phymatum* sp. nov., nodulate the roots of tropical legumes', *Systematic and Applied Microbiology*, vol 25, pp507–512
Vaneechoutte, M., Kamfer, P., de Baere, T., Falsen, E. and Verschraegen, G. (2004) '*Wautersia* gen. nov. sp. nov., a new genus accommodating the phylogenetic lineage including *Ralstonia eutropha* and related species, and proposal of *Ralstonia* [*Pseudomonas*] *syzygii* (Roberts et al, 1990) comb. nov.', *International Journal of Systematic and Evolutionary Microbiology*, vol 54, pp317–327
Velazquez, E., Igual, J. M., Willems, A., Fernandez, M. P., Munoz, E., Mateos, P. F., Abril, A., Toro, N., Normand, P., Cervantes, M., Gillis, M. and Martínez-Molina, E. (2001) '*Mesorhizobium chacoense* sp. nov., a novel species that nodulates *Prosopis alba* in the Chaco Arido region (Argentina)', *International Journal of Systematic and Evolutionary Microbiology*, vol 51, pp1011–1021
Vincent, J. M. (1970) *A Manual for the Practical Study of Root-Nodule Bacteria*, Blackwell Scientific Publications, Oxford
Vinuesa, P., Léon-Barrios, M., Silva, C., Willems, A., Jarabo-Lorenzo, A., Pérez-Galdona, R., Werner, D. and Martínez-Romero, E. (2005a) '*Bradyrhizobium canariense* sp. nov., an acid tolerant endosymbiont that nodulates endemic genistoid legumes (Papilionoideae: Genisteae) from Canary Islands, along with *Bradyrhizobium japonicum* bv. genistearum, *Bradyrhizobium* genospecies a and *Bradyrhizobium* genospecies b', *International Journal of Systematic and Evolutionary Microbiology*, vol 55, pp569–575
Vinuesa, P., Silva, C., Werner, D. and Martínez-Romero, E. (2005b) 'Population genetics and phylogenetic inference in bacterial molecular systematics: The roles of migration and recombination in *Bradyrhizobium* species cohesion and delineation', *Molecular Phylogenetics and Evolution*, vol 34, pp29–54

Wang, E. T., van Berkum, P., Beyene, D., Sui, X. H., Dorado, O., Chen, W. X. and Martínez-Romero, E. (1998) '*Rhizobium huautlense* sp. nov., a symbiont of *Sesbania herbacea* that has a close phylogenetic relationship with *Rhizobium galegae*', *International Journal of Systematic Bacteriology*, vol 48, pp687–699

Wang, E. T., van Berkum, P., Sui, X. H., Beyene, D., Chen, W. X. and Martínez-Romero, E. (1999b) 'Diversity of rhizobia associated with *Amorpha fruticosa* isolated from Chinese soils and description of *Mesorhizobium amorphae* sp. nov.', *International Journal of Systematic Bacteriology*, vol 49, pp51–65

Wang, E. T., Rogel-Hernández, A., Santos, A. G., Martínez-Romero, J., Cevallos, M. A. and Martínez-Romero, E. (1999a) '*Rhizobium etli* bv. mimosae, a novel biovar isolated from *Mimosa affinis*', *International Journal of Systematic Bacteriology*, vol 49, pp1479–1491

Wang, E. T., Tan, Z. Y., Willems, A., Fernández-López, M., Rinhold-Hurek, B. and Martínez-Romero, E. (2002) '*Sinorhizobium morelense* sp. nov., a *Leucena leucocephala*-associated bacterium that is highly resistant to multiple antibiotics', *International Journal of Systematic Microbiology*, vol 52, pp1687–1693

Wei, G. H., Wang, E. T., Tan, M. E., Zhu, M. E. and Chen, W. X. (2002) '*Rhizobium indigoferae* sp. nov. and *Sinorhizobium kummerowieae* sp. nov., respectively isolated from *Indigofera* spp. and *Kummerowia stipulacea*', *International Journal of Systematic and Evolutionary Microbiology,* vol 52, pp2231–2239

Wei, G. H., Wang, E. T., Zhu, M. E., Wang, E. T., Han, S. Z. and Chen, W. X. (2003) 'Characterization of rhizobia isolated from legume species within the genera *Astragalus* and *Lespedeza* grown in Loess Plateau region of China and description of *Rhizobium loessense* sp. nov.', *International Journal of Systematic and Evolutionary Microbiology*, vol 53, pp1575–1583

Willems, A., Fernández-Lopez, M., Muñoz-Adelantado, E., Goris, J., Vos, P., Martínez-Romero, E., Toro, N. and Gillis, M. (2003) 'Description of new *Ensifer* strains from nodules and proposal to transfer *Ensifer adhaerens* Cassida 1982 to *Sinorhizobium* as *Sinorhizobium adhaerens* comb. nov. Request for an opinion', *International Journal of Systematic and Evolutionary Microbiology*, vol 53, pp1207–1217

Wilson, K. H., Blitchington, R. B. and Green, R. C. (1990) 'Amplification of bacterial 16S ribosomal DNA with polymerase chain reaction', *Journal of Clinical Microbiology*, vol 28, pp1942–1946

Woomer, P. (1993) 'Most probable number counts of Rhizobia in soils', in J. M. Anderson and J. S. I. Ingram (eds) *Tropical Soil Biology and Fertility: A Handbook of Methods*, CAB International, Wallingford, pp172–178

Woomer, P., Bennet, J. and Yost, R. (1990) 'Agroclimatology and modeling – overcoming the inflexibility of most-probable-number procedures', *Agronomy Journal*, vol 82, pp349–353

Woomer, P., Singleton, P. W. and Bohlool, B. B. (1988a) 'Ecological indicators of native rhizobia in tropical soils', *Applied and Environmental Microbiology*, vol 54, pp1112–1116

Woomer, P., Singleton, P. W. and Bohlool, B. B. (1988b) 'Reliability of the most-probable technique for enumerating rhizobia in tropical soils', *Applied and Environmental Microbiology*, vol 54, pp1494–1497

Xu, L. M., Ge, C., Cui, Z., Li, J. and Fan, H. (1995) '*Bradyrhizobium liaoningense* sp. nov., isolated from the root nodules of soybeans', *International Journal of Systematic Bacteriology*, vol 45, pp706–711

Yao, Z. Y., Kan, F. L., Wang, E. T. and Chen, W. X. (2002) 'Characterization of rhizobia that nodulate legume species within the genus *Lespedeza* and description of *Bradyrhizobium yuanmigense* sp. nov.', *International Journal of Systematic Bacteriology*, vol 52, pp2219–2230

Young, J. M. (2003) 'The genus name *Ensifer* Casida 1982 takes priority over *Sinorhizobium* Chen et al 1988 and *Sinorhizobium morelense* Wang et al 2002 is a junior synonym of *Ensifer adhaerens* Casida 1982. Is the combination "*Sinorhizobium adhaerens*" (Casida 1982)

Willems et al 2002 legitimate? Request for an opinion', *International Journal of Systematic and Evolutionary Microbiology*, vol 53, pp2107–2110

Young, J. M., Kuykendall, L. D., Martínez-Romero, E., Kerr, A. and Sawada, H. (2001) 'A revision of *Rhizobium* Frank 1889, with an emended description of the genus, and the inclusion of all species of *Agrobacterium* Conn 1942 and *Allorhizobium undicola* de Lajudie et al, 1998 as new combinations: *Rhizobium radiobacter, R. rhizogenes, R. rubi, R. undicola* and *R. vitis*', *International Journal of Systematic Evolutionary Microbiology*, vol 51, pp89–103

Young J. P. W. and Haukka, K. E. (1996) 'Diversity and phylogeny of rhizobia', *New Phytologist*, vol 133, pp87–94

APPENDIX 6.1 COMPOSITION OF 79 MEDIUM FOR GROWTH OF LNB

79 medium (Fred and Waksman, 1928) (similar to YMA medium – Vincent,1970):

10g mannitol or sucrose
1mL sol. K_2HPO_4 (10%) (or 0.1 gL^{-1})
4mL sol. KH_2PO_4 (10%) (or 0.4 gL^{-1})
2mL sol. $MgSO_4 . 7H_2O$ (10%) (or 0.2 gL^{-1})
1mL sol. NaCl (10%) (or 0.1 gL^{-1})
100mL yeast extract (or 0.4 gL^{-1} powder)
5mL sol. 0.5% Bromothymol blue in 0.2 N KOH; make up to 1000 mL with distilled water pH 6.8–7.0.

Solid medium: 15g agar
Semi-solid medium: 1.75g agar

Autoclave at 120°C for 15 min;
if pH < 5.0 replace Bromothymol blue with Bromocresol green and increase agar to 20 gL^{-1}.

Appendix 6.2 Jensen's nutrient solution for growth of leguminous species in plastic pouches and Leonard jars

Components	Volume
K_2HPO_4 (2%)	10mL
$MgSO_4.7H_2O$ (2%)	10mL
NaCl (2%)	10mL
$CaHPO_4$ (10%)	10mL
$FeCl_3.6H_2O$ 1.7 % or $FeCl_3$ 1%	10mL
Micronutrient solution*	1mL
Distilled water (make up to)	1000mL

pH = 6.7, adjustment with KOH

Dilute solution to one fourth

* Micronutrient solution (for 1L of water).

H_3BO_3 .... 2.86g
$MnSO_4.4H_2O$ 2.03g
$ZnSO_4.7H_2O$ 0.22g
$CuSO_4.5H_2O$ 0.08g
$Na_2MoO_4.H_2O$ 0.09g

This is the same protocol for seedling agar. Vincent (1970) recommends the mixture for nutrient solution but also diluted as here.Then we use this solution with siratro in plastic pouches too.

Controls

N-controls are provided to a final concentration of approx 70ppm N (0.05% KNO_3 or NH_4NO_3). This may be added to the nutrient solution at the beginning of the experiment or later, i.e. 7–10 days after planting. If this is insufficient for the sustained growth and green colour of N-control plants towards the end of the experiment, it can be supplemented. However, higher concentrations (>0.07% KNO_3) can be toxic. Usually N-controls lack nodules as combined N inhibits nodulation. Thus, absolute controls, i.e., without soil inoculation, are necessary to check aseptic conditions of the experiment. Another important control, not mentioned by Vincent, is the use of an efficient strain as a positive control for nodulation and nitrogen fixation. If strains recommended as inoculants are available they should be used.

Chapter 7

Arbuscular Mycorrhizal Fungi (AMF)

Joseph D. Bagyaraj and Sidney L. Stürmer

INTRODUCTION

It is now very well documented that arbuscular mycorrhizal fungi (AMF) improve fitness and growth of plants that are important in agriculture, horticulture and forestry. The soil hyphal network produced by AMF during association with the plant host provides a greater absorptive surface than root hairs alone and thus increases significantly the absorption of relatively immobile ions such as phosphate, copper and zinc. In most tropical soils, available phosphorus is very low and thereby limiting for plant development. In addition, mycorrhizally infected plants have been shown to have greater tolerance to toxic metals, to root pathogens, to drought, to high soil temperature, to saline soils, to adverse soil pH and to transplant shock than non-mycorrhizal plants (Mosse et al, 1981; Bagyaraj, 1990; Bagyaraj and Varma, 1995). AMF have been reported from natural ecosystems such as deserts, sand dunes, tropical forests, salt marshes and managed systems such as pastures, orchards and field crops (Brundrett, 1991).

In the tropics, agriculture is practised in areas previously occupied by two main plant species-rich natural ecosystems: tropical forests and savannah woodlands. The conversion of these two ecosystems into agro-ecosystems, whether related to subsistence agriculture, production of cash crops or industrial forest plantations, provokes changes in the chemical, physical and biological characteristics of the edaphic environment. Sites are usually cleared of multispecies, uneven-aged vegetation and normally planted with a single species of one age-class. For AMF, conversion of natural ecosystems into distinct land use systems influences spore abundance and species composition. Jasper et al (1987) observed a drop in spore numbers and a shift in species composition after disturbance in some Australian sites. Similarly, Mason et al (1992) found that the number of spores of AMF in a plantation of *Terminalia ivoriensis* in Cameroon greatly decreased three months after complete clearance and also noticed a change in species composition. Johnson and Wedin (1997), however, found similar species richness in dry tropical forest and monodominant grassland, where 28 AMF morphotypes were detected with *Glomus aggregatum* and two undescribed *Glomus* species observed as the predominant fungi. Also, Picone (2000) demonstrated that spore density and the fungal community available for mycorrhizal formation were relatively similar between pasture and primary rain forest.

Measuring the taxonomic diversity of AMF has relied mostly on direct counting and the identification of field-recovered spores. However, this approach fails to detect cryptic species of AMF that are not sporulating at the sampling time (but are still associated with plant hosts) and hampers accurate identification of some species, especially those of the genera *Glomus* and *Gigaspora*. Trapping from field soil has been used successfully to detect and recover cryptic species during AMF diversity surveys in temperate region apple plantations (Miller et al, 1985) and grassland (Bever et al, 1996) and also in desert (Stutz and Morton, 1996). Although it is not commonly reported, measurement of taxonomic diversity should be accompanied by some assessment of the activity of the mycorrhizal community. Mycorrhizal infectivity is easily determined by the most probable number (MPN) method and can serve for comparative purposes (Porter, 1979). The mean infection percentage (MIP) assay (Moorman and Reeves, 1979) and the infection unit (IU) assay (Franson and Bethlenfalvay, 1989) are also available.

Taxonomy and classification of AMF

A taxonomic framework must be established prior to any biodiversity study, especially when comparisons among land use systems are required. Taxonomy and classification of AMF have changed radically over the last few years. Gerdemann and Trappe (1974) included all AMF species within the order Endogonales in the family Endogonaceae (Division Zygomycota). Morton and Benny (1990) transferred all AMF species to the order Glomerales (Division Zygomycota) with three families and six genera. Molecular and morphological data sets were used together to transfer members of the genus *Sclerocystis* to *Glomus* (Almeida and Schenck, 1990; Redecker et al, 2000) and to erect two new genera pertaining to two distinct families (Morton and Redecker, 2001). Schüßler et al (2001) moved all AMF species to a newly monophyletic division, the Glomeromycota, placing this group of organisms at the same level as the Basidiomycota and Ascomycota. They also included in the new classification scheme three new orders and several families. Recently, a new genus named *Pacispora* was proposed by Oehl and Sieverding (2004); spore formation in this genus is similar to that of *Glomus* but they differentiate inner germinal walls. This genus is included within the order Diversisporales in the molecular phylogeny proposed by Schüßler et al (2001). Although molecular techniques have been applied to clarify phylogenetic classification among AMF species, identification of species has been based mainly on morphological characteristics of spores. Therefore, we propose a classification scheme to be followed for diversity surveys of AMF, merging the schemes from Morton and Benny (1990), with those of Morton and Redecker (2001) and Schüßler et al (2001) (Table 7.1).

Table 7.1 *Taxonomic framework proposed for diversity studies of arbuscular mycorrhizal fungi and morphological characters defining genera in Glomerales*

Division Glomeromycota Schüßler, Scharzott and Walker

Order Glomerales Morton and Benny

Suborder Glomineae Morton and Benny

Family Glomeraceae Pirozysnki and Dalpé

Glomus Tulasne and Tulasne (85 species)

Spores formed blastically on a subtending hypha, singly, in loose aggregates or in a sporocarp. Vesicles are thin walled and ellipsoid. Intraradical hyphae rarely coiled, with cross-connecting branch hyphae. Mycorrhiza stains darkly. Arbuscules with flared or cylindrical trunks with incremental narrowing of branch hyphae. Spores with spore wall formed by a variable number of layers all originating from the subtending hyphae, no germinal walls differentiated. Germination through the lumen of the subtending hyphae or through the spore wall.

Family Acaulosporaceae Morton and Benny

Acaulospora Gerd. and Trappe emend. Berch (31 species)

Spores formed laterally from the neck of a sporiferous saccule which leaves one scar on the spore surface. Vesicles vary in shape with knobs and concavities. Intraradical hyphae straight or coiled near the entry points. Mycorrhiza stains weakly. Arbuscules with flared or cylindrical trunks with incremental narrowing of branch hyphae. Spores with spore wall formed by three layers and two inner germinal walls, each with two thin layers that can be adherent. The innermost germinal wall has a beaded surface. Germination through a flexible, plate-like germination orb.

Entrophospora Ames and Schneider (four species)

Spores formed within the neck of a sporiferous saccule which leaves two scars on the spore surface. Vesicles, arbuscules, intraradical hyphae and mycorrhizae staining as in *Acaulospora*. Spores with spore wall formed by two layers. Other spore sub-cellular structures and germination identical to that in *Acaulospora*.

Family Archaeosporaceae Morton and Redecker

Archaeospora Morton and Redecker (three species)

Spores formed terminally from subtending hyphae or as a branch from a structure resembling a sporiferous saccule. Arbuscules and intraradical hyphae stain lightly. Vesicles and auxiliary cells are not differentiated. Spores with spore wall formed by three to four layers and no true bi-layered germinal wall formed. Dimorphic species found forming acaulosporoid and glomoid spores.

Family Paraglomeraceae Morton and Redecker

Paraglomus Morton and Redecker (two species)

Spores formed terminally from a subtending hypha (as in *Glomus*). Arbuscules and intraradical hyphae stain lightly. Vesicles and auxiliary cells are not differentiated. Spore sub-cellular structures and germination as in *Glomus*.

Family Pacisporaceae Walker, Blaszkowski, Schüßler and Schwarzott

Pacispora Sieverding and Oehl (seven species)

Spores formed terminally from a subtending hypha (as in *Glomus*). Spore wall usually formed by three distinct layers and germinal wall composed also of three layers. The second layer of the germinal wall usually reacts to Melzer's reagent. Walls of subtending hyphae continuous with the first and second layer of the spore wall. Spore germination from the germinal wall directly through the spore wall.

Suborder Gigasporineae Morton and Benny

Family Gigasporaceae Morton and Benny

Gigaspora Gerd. and Trappe (five species)

Spores formed terminally on a bulbous sporogenous cell; auxiliary cells finely papillate or echinulate. No vesicles produced. Intraradical hyphae frequently coiled, especially near entry points, often knobby or with projections. Arbuscules with swollen trunks with abrupt narrowing of branch hyphae. Spores with spore wall formed by two permanent layers, no inner germinal walls differentiated. At germination, a thin layer interspersed with warts differentiates and germ tube grows throughout the spore wall.

Scutellospora Walker and Sanders (30 species)

Spores formed terminally on a bulbous sporogenous cell; auxiliary cells almost smooth to knobby. No vesicles produced. Arbuscules and intraradical hyphae similar in morphology to *Gigaspora*. Spores with spore wall formed by two permanent layers and one to three inner germinal walls, each with two layers. Germ tube grows from flexible, plate-like germination shield that differentiates on the surface of the last germinal wall.

Source: Compiled from Morton and Benny, 1990; Oehl and Sieverding, 2004 and http://invam.caf.wvu.edu

METHODS TO ASSESS AMF INFECTIVE PROPAGULES AND COLONIZATION

Most probable number (MPN)

The sampling protocol proposed to estimate AMF diversity (described in Chapter 2) is the same as that followed for all microbes and will not be further discussed here. The most probable number method has been used to estimate the number of infective propagules of AMF in several soils. It is based on a series of tenfold soil dilutions where presence or absence of mycorrhizal colonization is recorded and results given as a probability of the number of infective propagules based on a statistical table. One of the disadvantages is that the preparation of the assay is time-consuming; the number calculated has a high 95 per cent confidence level though (Adelman and Morton, 1986). The advantage is that the analysis of trap host roots to check for colonization is easy and fast to carry out for someone familiar with mycorrhizal morphology. As the result is a single number, the method is appropriate for comparisons among land use systems if MPN assays are set up simultaneously.

Materials needed

- plant tubes (15×2.5cm) and racks;
- plastic bags (30×20cm);
- sterilized diluent (sand : soil mix 1:1);
- onion seeds.

The recommended host plant for the MPN assay is onion because it is dependent on mycorrhiza and its roots are easy to stain. However, any other host can be used if they have some dependence on the mycorrhizal condition. If onion seeds are not available, C4 grasses like *Sorghum* and *Paspalum* can be used.

Procedure

1. Weigh 30g of field soil in a plastic bag and add to it 270g of sterilized diluents. Shake thoroughly to get a tenfold dilution.
2. Remove 30g from the 10^1 dilution and place it into another bag containing 270g of sterilized diluents. Shake thoroughly to get a hundredfold dilution. Make further tenfold dilutions up to 10^4 dilution (or higher dilution if needed).
3. Distribute soil from each dilution into plant tubes; use five replicate tubes per dilution.
4. Sow seeds of onion into each tube; if onion seeds are not available, any other suitable host can be used (see above).
5. After emergence, trim to only one plant per tube and let plants grow in a greenhouse or growth room for six weeks.
6. At harvest, wash soil from the roots and stain them with trypan blue (see below).
7. Under a dissecting microscope, determine the presence or absence of mycorrhizal colonization in each replicate. Counts of positive tubes (those containing mycorrhiza) in different dilutions are used to calculate MPN values. Use tables of Cochran (1950), Fisher and Yates (1963) or Alexander (1965).

The procedure for the calculation of the MPN is illustrated with the following example. Let's assume that, considering the five replicates (tubes) for each of the four dilutions (10^1, 10^2, 10^3 and 10^4), the following sequence of numbers of positive tubes is obtained: 5, 5, 3, 2. This means that all five replicate tubes are positive for mycorrhizal colonization at 10^1 and 10^2 dilutions, that three replicate tubes are positive at 10^3 dilution and two replicate tubes are positive at 10^4 dilution. For the calculation of MPN, only three numbers of a given sequence are required. The first number (N1) specifies the number of tubes that are positive for mycorrhizal colonization, in that dilution series that shows the highest number of positive tubes (the less concentrated series is taken, if the number of positive tubes is the same in subsequent dilutions). The two other numbers (N2 and N3) are those corresponding to the next two dilutions. In this example the combination would be:

N1 = 5, N2 = 3 and N3 = 2.

Using the MPN table, the value given for this combination of positive values is 1.4. To obtain the MPN of infective propagules of AMF in the sample, this value has to be multiplied by the middle dilution (in this case 10^3). Therefore, the soil has 1.4×10^3 infective propagules g^{-1}.

Staining roots to observe mycorrhizal colonization

Colonization of root cortex cells by AMF does not alter root morphology in contrast to ectomycorrhizal associations. Therefore, to detect and measure mycorrhizal colonization, roots are subjected to a clearing and staining procedure. The method most extensively used is that described by Philips and Hayman (1970). Kormanik and McGraw (1982) eliminated phenol from the staining and destaining solutions and Koske and Gemma (1989) modified the procedure by eliminating lactic acid from these solutions, without interfering with the intensity of staining. Considering the cost of chemicals and the economy of attempting to reduce the number of chemicals without interfering with the quality of staining, the method of Philips and Hayman (1970) modified by Koske and Gemma (1989) is proposed for staining roots:

Materials needed

- 10 per cent KOH solution (potassium hydroxide);
- 1 per cent HCl solution (hydrochloric acid);
- acidic glycerol solution (500mL glycerol, 450mL H_2O, 50mL 1 per cent HCl);
- 0.05 per cent trypan blue in acidic glycerol solution (0.5g/L);
- alkaline H_2O_2 solution (3mL 20 per cent NH_4OH, 30mL 3 per cent H_2O_2, 567mL tap water).

Procedure

1 Wash roots free from soil debris and rinse in several changes of tap water.
2 Soak roots in 10 per cent KOH at 90°C for one hour or at 120°C for 15 minutes.
3 Remove KOH and rinse roots with water (two to three times) to remove excess KOH.
4 If roots are too pigmented, soak in alkaline H_2O_2 solution for 10–30 minutes. Rinse roots with water again after this step.

5 Soak roots in 1 per cent HCl solution for five minutes.
6 Remove HCl. *Do not rinse* roots after this step as they must be acidified for proper staining.
7 Stain roots in acidic glycerol solution containing trypan blue at 90°C for one hour or at 120°C for five minutes.
8 Discard stain solution and keep roots in acidic glycerol (without trypan blue) or water at room temperature or 4°C.

Comments: acidic glycerol solution can be replaced by a lacto-glycerol solution (20mL lactic acid, 40mL glycerol and 40mL distilled water) if desired. Step number 4 can be omitted if roots are not highly pigmented; alkaline H_2O_2 solution must be prepared fresh just before use. For long-term storage, roots can be kept in water to which a few drops of 0.1 per cent sodium azide have been added. For steps 2 and 7, a water bath is suitable to keep root samples at 90°C.

Measuring mycorrhizal colonization

Measurement of root mycorrhizal colonization is frequently carried out on field-collected roots (directly from the soil or from individual plants) or on experimental plants grown under greenhouse conditions. It estimates growth of a fungal isolate or a fungal community within the root cortex. The grid-line intersect method (Giovannetti and Mosse, 1980), which is presented below, is commonly used to measure mycorrhizal root length and percentage of mycorrhizal colonization.

Materials needed

- Petri dish with a 1.1×1.1cm grid on the bottom;
- dissecting needles.

Procedure

1 Spread stained roots out evenly in the Petri dish.
2 Under a dissecting microscope, scan vertical and horizontal lines of the grid.
3 Record: a) the total number of intersections of roots and grid lines and b) the number of intersections with mycorrhizally infected roots.
4 Calculate the percentage of mycorrhizal colonization (%MycCol) using the formula below:

%MycCol = ('Total no of intersections with mycorrhizally infected roots'/'Total no of intersections between root and the gridline') ×100

Comments: the data obtained at step 3 can also be used to determine plant root and mycorrhizal root lengths. If the entire root system is spread out in the Petri dish, with the grid lines separated at 1.1cm, the total number of roots intersecting the grid lines represents the total root length in cm. Therefore the total mycorrhizal root length will be the number of intersections having mycorrhizal structures. If a small sample of roots is taken from the entire root, then stained and measured in the microscope by this method, it is possible to calculate the total root length and the mycorrhizal root length of the entire plant, making use of the dry weight ratio between plant and subsample. For further details see http://invam.caf.wvu.edu/methods/mycorrhizae/rootlengths.htm.

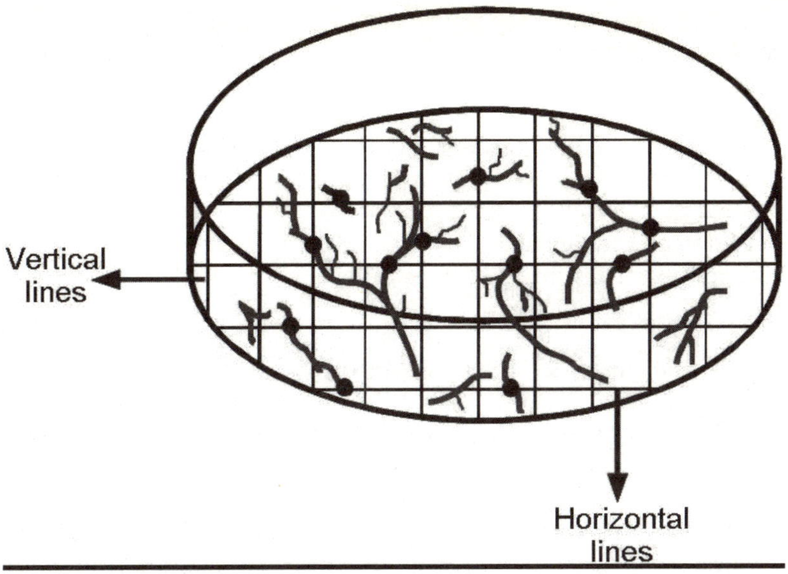

Note: In the diagram mycorrhizal infections are represented by black dots.

Figure 7.1 *Petri dish with stained roots evenly distributed to score mycorrhizal colonization by a line intersection method*

Figure 7.1 provides an illustration of the grid intersection methods. In this example, after spreading the roots, the number of intersections between a root piece and the horizontal and vertical lines are as follows:

25 intersects (horizontal lines)
18 intersects (vertical lines) and
12 intersects (black dots) presenting mycorrhizal colonization.

Therefore, the *Root Length* = 43cm (25+18) and the *Mycorrhizal Root Length* = 12cm (six positive intersects on horizontal lines and six positive intersects on vertical lines).

The *Percentage of Mycorrhizal Colonization* = (12/43) ×100 = 27.9.

METHODS TO ASSESS AMF DIVERSITY

Spore extraction from the field

Identification of AMF species is based on the analysis of sub-cellular structures of asexual spores. Morphology of internal and external mycelium during mycorrhizal association is practically indistinguishable among species within the same genus and between genera. Spore sub-cellular structures, by contrast, are highly conserved and phenotypically stable regardless of environment and plant host (Morton et al, 1995). Thus, spores are the only part of the fungal organism that can be used to delimit species.

Spores are extracted from the soil by wet sieving (Gerdemann and Nicolson, 1963) followed by sucrose gradient centrifugation.

Materials needed

- set of nested sieves, containing at least a 710µm and a 45µm mesh;
- plastic bucket or a large beaker;
- beakers, Petri dishes, watch glass;
- centrifuge tubes and centrifuge;
- sucrose solutions (20 per cent and 60 per cent).

Procedure

1. Place 100g of soil sample in the bucket/beaker and suspend it in at least 500mL water.
2. Stir thoroughly to bring spores into suspension and let stand for 30 seconds.
3. Pass the suspension thorough two nested sieves with 710µm and 45µm mesh successively.
4. Place the material retained in the 710µm sieve into a large Petri dish and observe under the dissecting microscope for large spores (*Gigaspora* and *Scutellospora* spores) and sporocarps.
5. Place the material retained in the 45µm sieve into a beaker with a small amount of water; then transfer this material to a centrifuge tube containing a 20–60 per cent sucrose gradient.
6. Centrifuge at 2000rpm for one minute.
7. Drop the supernatant back into the 45µm sieve and wash with tap water to remove excess of sucrose.
8. Transfer spores and debris from the 45µm sieve to a Petri dish and under the microscope collect the spores on a clean watch glass.

Comments: for step 3, other sieves can be stacked between the two recommended ones. This will separate spores by size but will increase the amount of work as material retained in each sieve must be centrifuged separately. The sieves recommended (with 710µm and 45µm mesh) are suitable to recover most AMF species as spore size for the majority of species ranges from 40 to 600µm. However, if there are too many root pieces and other debris it is preferable to pour the soil suspension through a 1mm sieve before passing it through the 710µm sieve. Sucrose solution is made from commercial sugar bought in the supermarket. To prepare the sucrose gradient, add 15mL of the 20 per cent sucrose solution in a 50ml centrifuge tube and then, on the bottom of it, add another 15mL of the 60 per cent solution.

Spores can be separated from organic debris and collected from the Petri dish by using either a fine forceps or an extruded glass Pasteur pipette. At this time, spores can be separated by morphotypes according to colour and size and if the spores are in good health they can be identified up to genus and species level. Soils with high content of clay might clog the sieves; a small amount of 0.1M sodium pyrophosphate (Fogel and Hunt, 1979) can be added to disperse clay particles and free up the sieve openings.

Species identification and mounting slides

In most cases, spores extracted from the field are not identifiable under the dissecting microscope and therefore need to be mounted on slides and observed under a compound microscope. Some comprehensive identification keys for AMF (e.g. Hall and Fish, 1979; Trappe, 1982) were published before some of the important recent taxonomic revisions. Koske and Walker (1985) proposed a key for some *Scutellospora* species with ornamented spore walls and Bentivenga and Morton (1995) proposed a key to identify the five species of *Gigaspora*. Identification of AMF species can be done by comparison with original species description, referring to the *Manual for the Identification of VAM Fungi* (Schenck and Pérez, 1990) and comparisons with species descriptions and illustrations provided by Dr Joseph Morton at the INVAM (International Culture Collection of Arbuscular Mycorrhizal Fungi, West Virginia University, WV, USA) website (http://invam.caf.wvu.edu) and by Dr Janusz Blaszkowski (Department of Plant Pathology, Agricultural University of Szczecin, Poland) at www.agro.ar.szczecin.pl/~jblaszkowski. The Schenck and Pérez manual is a copy of the original species descriptions and therefore it harbours all the problems inherent in these descriptions (lack of standardization on spore sub-cellular structures, lack of good photographs for comparisons and description from field-collected spores which might not include all important taxonomic characteristics). Nevertheless, this manual incorporates in one volume all species description up to 1990. On the other hand, the INVAM and Blaszkowski's sites possess excellent pictures with details of spore sub-cellular structure, spore size and a standardized description of AMF species; on these websites 79 and 55 species are described respectively out of 160 species of AMF described in total. Species names and authorities are provided for reference in Appendices 7.1, 7.2, 7.3 and 7.4.

Pictures of spores of representative species of different genera are shown in Plates 4 and 5. Plate 4 depicts spores and structures produced by AMF species of the family Gigasporaceae (pictures a, b, c and d) and Acaulosporaceae (e, f, g, and h). Picture 'a' shows a spore of *Gigaspora albida* with the suspensor bulbous cell typical of this family, 'b' a spore of *Scutellospora scutata* with suspensor bulbous cell (notice round brown germination shield contrasting with the hyaline spore colour), 'c' a detail of the spore wall ornamentation (warts) of *Scutellospora coralloidea*, and 'd' the knobby auxiliary cells differentiated by members of *Scutellospora*. A spore of *Entrophospora colombiana* (family *Acaulosporaceae*) with spore wall, germinal wall 1 (gw1) and germinal wall 2 (gw2) with its innermost layer reacting to Melzer's reagent is depicted in picture 'e'. A spore of *Entrophospora colombiana* showing the two characteristic scars is seen in 'f'. Picture 'g' represents a spore of *Acaulospora scrobiculata* showing the scar that is left on the spore after the sporiferous saccule detached, and spores of *Acaulospora* sp. showing some sporiferous saccules attached to the spores is seen in the last picture, 'h'.

Plate 5 presents spores and structures produced by species of the family Glomeraceae (a, b, c and d), Archaeosporaceae (e and f) and Paraglomeraceae (g and h). Picture 'a': a spore of *Glomus clarum* indicating the subtending hypha (note that the innermost layer of the spore wall detaches and looks similar to a germinal wall; picture 'b' shows the subtending hypha wall continuous with the spore wall of a spore of a *Glomus* sp.; a sporocarp of *Glomus clavispora* in picture 'c' and a sporocarp of *Glomus* sp. in 'd'. Picture 'e' depicts a spore of *Archaeospora leptoticha* with a sporiferous saccule, picture 'f' provides a detail of *Archaeospora leptoticha* showing protuberances and depressions of layers 2 and 3 of the spore wall (indicated with arrows), picture 'g' exemplifies the spore wall structure formed by three layers (L1, L2 and L3) of the *Paraglomus occultum*, and a spore of *Paraglomus brasilianum*, likewise with a three-layered spore wall (L1, L2 and L3) is shown in picture 'h' (note that L2 is ornamented with minute ridges).

Taxonomically important characteristics that can be observed under the dissecting microscope are spore size, colour and shape, presence of subtending hyphae, presence of suspensor cells and presence of the sporiferous saccule (rarely seen from field collected spores). Under the compound microscope, some important taxonomic characteristics to be observed are the presence and type of ornamentation on the spore wall, Melzer's reaction, number of germinal walls, and thickness of spore wall, amongst others. Spores need to be mounted in permanent mounting media like PVLG (polyvinyl-lacto-glycerol) and PVLG mixed with Melzer's reagent.

Materials needed

- glass slides, coverslips and labels;
- dissecting needle;
- PVLG solution (100mL distilled water, 100mL lactic acid, 10mL glycerol, 16.6g polyvinyl alcohol (PVA);
- Melzer's reagent (100g chloral hydrate, 100mL distilled water, 1.5g iodine, 5g potassium iodide). Mix Melzer's reagent with PVLG (1:1) to make a PVLG + Melzer solution.

Procedure

1. On a glass slide, add one drop of PVLG and one drop of PVLG + Melzer solution.
2. Add spores to the centre of each drop by using a fine forceps or extruded micropipette (in this case, be careful not to add too much water).
3. With a dissecting needle, mix spores with the PVLG and PVLG + Melzer solution and let stand for at least five minutes to dry the surface of the drop slightly.
4. Gently place a coverslip over the drop of PVLG and another coverslip over the drop of PVLG + Melzer.
5. Using a dissecting needle, crush each spore individually under the dissecting microscope. This step is important to expose germinal walls and their layers.
6. Label the slides including sample number, genus and species (if known) and date.
7. Keep slides at room temperature for five days and then incubate them at 40–60°C for two days to harden the mountant medium.

Comments: step 2 is crucial for mounting spores on a slide – if too much water is added, spores will slide out to the edges when the cover slip is placed on the top of each drop.

Establishment of trap cultures

Trap cultures are more frequently used in studies on AMF diversity because they reveal species that are not sporulating in the field at time of sampling. Also, they provide fresh and healthy spores which can be used to establish single isolate cultures of AMF. The method proposed is based on the protocols of Stutz and Morton (1996).

Materials needed

- sterile sand;
- 1.5kg plastic pots;
- plastic trays or plastic bags;
- seeds of *Sorghum sudanense* (sudangrass) and *Vigna unguiculata* (cowpea).

Procedure

1. On a plastic tray (or in a plastic bag), homogenize the soil from the field with sterile sand (50 per cent field soil and 50 per cent sand).
2. Place this mixture in 1.5kg plastic pots and seed heavily with a mixture of sudangrass and cowpea (40–50 seeds per pot). Cover seeds with the soil/sand mixture.
3. After three to four months under greenhouse conditions, sample one or two 50mL soil cores from the pot, extract spores and identify as explained above.

Comments: it is important that, during sampling of field soil, plant roots are included as they also serve as propagules to start trap cultures. In step 2, other plant hosts are suitable as well if sudangrass and cowpea are not available; it is recommended that any C4 grass and any other legume be used as hosts. Bever et al (1996) used a different procedure to establish trap cultures, called 'transplant trap cultures', where intact field-collected plants are transplanted to a pot with a substrate free of AMF and sporulation evaluated after three to four months.

References

Adelman, M. J. and Morton, J. B. (1986) 'Infectivity of vesicular-arbuscular mycorrhizal fungi: Influence of host-soil diluent combinations on MPN estimates and percentage colonization', *Soil Biology and Biochemistry*, vol 18, pp7–13

Alexander, M. (1965) 'Most-probable-number method for microbial populations', in A. Klute (ed) *Methods of Soil Analysis, Part 2: Chemical and Microbiological Methods*, Soil Science Society of America, Madison, WI, pp1467–1472

Almeida, R. T. and Schenck, N. T. (1990) 'A Revision of the Genus *Sclerocystis* (Glomaceae, Glomales)', *Mycologia*, vol 82, pp703–714

Bagyaraj, D. J. (1990) 'Ecology of vesicular-arbuscular mycorrhizae', in D. K. Arora, B. Rai, K. G. Mukerjii and G. R. Knudsen (eds) *Handbook of Applied Mycology: Soil and Plants*, Marcel Dekker Inc, New York

Bagyaraj, D. J. and Varma, A. (1995) 'Interaction between arbuscular mycorrhizal fungi and plants: Their importance in sustainable agriculture and in arid and semiarid tropics', *Advances in Microbial Ecology*, vol 14, pp119–142

Bentivenga, S. P. and Morton, J. B. (1995) 'A monograph of the genus *Gigaspora*, incorporating developmental patterns of morphological characters', *Mycologia*, vol 87, pp720–732

Bever, J. D., Morton J., Antonovics, J. and Schultz, P. A. (1996) 'Host-dependent sporulation and species diversity of arbuscular mycorrhizal fungi in a mown grassland', *Journal of Ecology*, vol 84, pp71–82

Brundrett, M. C. (1991) 'Mycorrhizas in natural ecosystems', *Advances in Ecological Research*, vol 21, pp171–213

Cochran, W. G. (1950) 'Present status of biometry', *Biometrics*, vol 6, pp75–78

Fisher, R. A. and Yates, F. (eds) (1963) *Statistical Tables for Biological, Agricultural and Medical Research*, Oliver and Boyd, Edinburgh

Fogel, R. and Hunt, G. (1979) 'Fungal and arboreal biomass in a Western Oregon Douglas-fir ecosystem: Distribution pattern and turnover', *Canadian Journal of Forest Research*, vol 9, pp245–256

Franson, R. L. and Bethlenfalvay, G. J. (1989) 'Infection unit method of vesicular-arbuscular mycorrhizal propagule determination', *Soil Science Society of America Journal*, vol 53, pp754–756

Gerdemann, J. W. and Nicolson, T. H. (1963) 'Spores of mycorrhizal *Endogone* species extracted from soil by wet sieving and decanting', *Transactions of the British Mycological Society*, vol 46, pp235–244

Gerdemann, J. W. and Trappe, J. M. (1974) 'Endogonaceae in the Pacific Northwest', *Mycologia Memoir*, vol 5, pp1–76

Giovannetti, M. and Mosse, B. (1980) 'An evaluation of techniques for measuring vesicular arbuscular mycorrhizal infection in roots', *New Phytologist*, vol 84, pp489–500

Hall, I. R. and Fish, B. J. (1979) 'A key to the Endogonaceae', *Transactions of the British Mycological Society*, vol 73, pp261–270

Jasper, D. A., Robson, A. D. and Abbott, L. K. (1987) 'The effect of surface mining on the infectivity of vesicular-arbuscular mycorrhizal fungi', *Australian Journal of Botany*, vol 35, pp641–652

Johnson, N. C. and Wedin, D. A. (1997) 'Soil carbon, nutrients and mycorrhizae during conversion of dry tropical forest to grassland', *Ecological Applications*, vol 7, pp171–182

Kormanik, P. P. and McGraw, A. C. (1982) 'Quantification of vesicular-arbuscular mycorrhizae in plant roots', in N. C. Schenck (ed) *Methods and Principles of Mycorrhizal Research*, American Phytopathological Society, St Paul, MN

Koske, R. E. and Gemma, J. N. (1989) 'A modified procedure for staining roots to detect VA mycorrhizas', *Mycological Research*, vol 92, pp486–505

Koske, R. E. and Walker, C. (1985) 'Species of *Gigaspora* (Endogonaceae) with roughened outer walls', *Mycologia*, vol 77, pp702–720

Mason, P. A., Musoko, M. O. and Last, F. T. (1992) 'Short-term changes in vesicular-arbuscular mycorrhizal spore populations in *Terminalia* plantations in Cameroon', in D. J. Read, D. H. Lewis, A. H. Fitter and I. J. Alexander (eds) *Mycorrhizas in Ecosystems*, CAB International Wallingford, pp261–267

Miller, D. D., Domoto, P. A. and Walker, C. (1985) 'Mycorrhizal fungi at eighteen apple rootstocks plantings in the United States', *New Phytologist*, vol 100, pp379–391

Mosse, B., Stribley, D. P. and Le Tacon, E. (1981) 'Ecology of mycorrhizae and mycorrhizal fungi', *Advances in Microbial Ecology*, vol 5, pp137–210

Moorman, T. and Reeves, F. B. (1979) 'The role of endomycorrhizae in revegetation practices in the semi-arid West, II: A bioassay to determine the effect of land disturbance on endomycorrhizal populations', *American Journal of Botany*, vol 66, pp14–18

Morton, J. B. and Benny, G. L. (1990) 'Revised classification of arbuscular mycorrhizal fungi (Zygomycetes): A new order, Glomales, two new suborders, Glomineae and Gigasporineae, and two new families, Acaulosporaceae and Gigasporaceae, with an emendation of Glomaceae', *Mycotaxon*, vol 37, pp471–491

Morton, J. B. and Redecker, D. (2001) 'Two new families of Glomales, Archaeosporaceae and Paraglomaceae, with two new genera *Archaeospora* and *Paraglomus*, based on concordant molecular and morphological characters', *Mycologia*, vol 93, pp181–195

Morton, J. B., Bentivenga, S. P. and Bever, J. D. (1995) 'Discovery, measurement, and interpretation of diversity in arbuscular endomycorrhizal fungi (Glomales, Zygomycetes)', *Canadian Journal of Botany*, vol 73 (suppl 1), ppS25–S32

Oehl, F. and Sieverding E. (2004) '*Pacispora*, a new vesicular-arbuscular mycorrhizal fungal genus in the Glomeromycetes', *Journal of Applied Botany*, vol 78, pp72–82

Philips, J. M. and Hayman, D. S. (1970) 'Improved procedures for clearing roots and staining parasitic and vesicular-arbuscular mycorrhizal fungi for rapid assessment of infection', *Transactions of the British Mycological Society*, vol 55, pp158–161

Picone, C. (2000) 'Diversity and abundance of arbuscular-mycorrhizal fungus spores in tropical forest and pasture', *Biotropica*, vol 32, pp734–750

Porter, W. M. (1979) 'Most probable number method for enumerating infective propagules of vesicular arbuscular mycorrhizal fungi in soil', *Australian Journal of Soil Research*, vol 17, pp515–519

Redecker, D., Morton, J. B. and Bruns, T. D. (2000) 'Molecular phylogeny of the arbuscular mycorrhizal fungi *Glomus sinuosum* and *Sclerocystis coremioides*', *Mycologia*, vol 92, pp282–285

Schenck, N. C. and Pérez, Y. (1990) *Manual for Identification of VA Mycorrhizal Fungi*, 3rd edition, Synergistic Publications, Gainesville

Schüßler, A., Schwarzott, D. and Walker, C. (2001) 'A new fungal phylum, the Glomeromycota: Phylogeny and evolution', *Mycological Research*, vol 105, pp1413–1421

Stutz, J. C. and Morton, J. B. (1996) 'Successive pot cultures reveal high species richness of arbuscular endomycorrhizal fungi in arid ecosystems', *Canadian Journal of Botany*, vol 74, pp1883–1889

Trappe, J. M. (1982) 'Synoptic key to the genera and species of Zygomycetous mycorrhizal fungi', *Phytopathology*, vol 72, pp1102–1108

Appendix 7.1 AMF species of the family Glomeraceae, genus Glomus

Genus/*Species*

Glomus

G. aggregatum Schenck and Smith
G. albidum Walker and Rhodes
G. ambisporum Smith and Schenck
G. antarticum Cabello
G. arborense McGee
G. aurantium Blaszkowski, Blanke, Renker and Buscot
G. australe (Berkeley) Berch
G. boreale (Thaxter) Trappe and Gerd.
G. botryoides Rothwell and Victor
G. caledonium (Nicol. and Gerd.) Trappe and Gerd.
G. callosum Sieverding
G. canadense (Thaxter) Trappe and Gerd.
G. cerebriforme McGee
G. citricola Tang and Zang
G. claroideum Schenck and Smith
G. clarum Nicol. and Schenck
G. clavisporum (Trappe) Almeida and Schenck
G. constrictum Trappe
G. convolutum Gerd. and Trappe
G. coremioides (Berk. and Broome) Redecker and Morton
G. coronatum Giovannetti
G. corymbiforme Blaszkowski
G. delhiense Mukerji, Bhattacharjee and Tewari
G. deserticola Trappe, Bloss and Menge
G. diaphanum Morton and Walker
G. dimorphicum Boyetchko and Tewari
G. dominikii Blaszkowski
G. etunicatum Becker and Gerdemann
G. fasciculatum (Thaxter) Gerd. and Trappe emend. Walker and Koske
G. flavisporum (Lange and Lund) Trappe and Gerd.
G. formosanum Wu and Chen
G. fragile (Berk. and Broome) Trappe and Gerd.
G. fragilistratum Skou and Jakobsen
G. fuegianum (Spegazzini) Trappe and Gerdemann
G. fulvum (Berk. and Broome) Trappe and Gerd.
G. geosporum (Nicol. and Gerd.) Walker
G. globiferum Koske and Walker
G. glomerulatum Sieverding
G. halonatum Rose and Trappe
G. heterosporum Smith and Schenck
G. hoi Berch and Trappe
G. insculptum Blaszkowski
G. intraradices Schenck and Smith
G. invermaium Hall
G. laccatum Blaszkowski
G. lacteum Rose and Trappe
G. liquidambaris (Wu and Chen) Almeida and Schenck
G. macrocarpum Tulasne and Tulasne
G. maculosum Miller and Walker
G. magnicaule Hall
G. manihotis Howeler, Sieverding and Schenck
G. melanosporum Gerd. and Trappe
G. microaggregatum Koske, Gemma and Olexia
G. microcarpum Tulasne and Tulasne
G. minutum Blaszk., Tadych and Madej
G. monosporum Gerdemann and Trappe
G. mortonii Bentivenga and Hetrick
G. mosseae (Nicol. and Gerd.) Gerd. and Trappe
G. multicaule Gerd. and Bakshi
G. multisubstensum Mukerji, Bhattacharjee and Tewari
G. nanolumen Koske and Gemma
G. pallidum Hall
G. pansihalos Berch and Koske
G. proliferum Dalpe and Declerck
G. przelewicensis Blaszkowski
G. pubescens (Sacc. and Ellis) Trappe and Gerdemann
G. pulvinatum (Henn.) Trappe and Gerdemann
G. pustulatum Koske, Friese, Walker and Dalpe
G. radiatum (Thaxter) Trappe and Gerd.
G. reticulatum Bhattacharjee and Mukerji
G. rubiforme (Gerd. and Trappe) Almeida and Schenck
G. segmentatum Trappe, Spooner and Ivory
G. sinuosum (Gerd. and Bakshi) Almeida and Schenck
G. sterilum Mehrotra and Baijal
G. taiwanense (Wu and Chen) Almeida and Schenck
G. tenebrosum (Thaxter) Berch
G. tenerum Tandy emend. McGee
G. tenue (Greenhall) Hall
G. tortuosum Schenck and Smith
G. trimurales Koske and Halvorson
G. tubiforme Tandy
G. versiforme (Karsten) Berch
G. vesiculiferum (Thaxter) Gerd. and Trappe
G. viscosum Nicol.
G. xanthium Blaszk., Blanke, Renker and Buscot
G. warcupii McGee

Source: Compiled from http://invam.caf.wvu.edu

Appendix 7.2 AMF species of the family Acaulosporaceae, genera Acaulospora and Entrophospora

Genus/*Species*	
Acaulospora	*A. polonica* Blaszkowski
A. bireticulata Rothwell and Trappe	*A. rehmii* Sieverding and Toro
A. capsicula Blaszkowski	*A. rugosa* Morton
A. cavernata Blaszkowski	*A. scrobiculata* Trappe
A. delicata Walker, Pfeiffer and Bloss	*A. spinosa* Walker and Trappe
A. denticulata Sieverding and Toro	*A. splendida* Sieverding, Chaverri and Rojas
A. dilatata Morton	*A. sporocarpia* Berch
A. elegans Trappe and Gerdemann	*A. taiwania* Hu
A. excavata Ingleby and Walker	*A. thomii* Blaszkowski
A. foveata Trappe and Janos	*A. tuberculata* Janos and Trappe
A. gedanensis Blaszkowski	*A. undulata* Sieverding
A. koskei Blaszkowski	*A. walkeri* Kramadibrata and Hedger
A. lacunosa Morton	
A. laevis Gerdemann and Trappe	*Entrophospora*
A. longula Spain and Schenck	*E. baltica* Blaszkowski
A. mellea Spain and Schenck	*E. colombiana* Spain and Schenck
A. morrowiae Spain and Schenck	*E. infrequens* (Hall) Ames and Schneider
A. myriocarpa Spain, Sieverding and Schenck	*E. kentinensis* Wu and Liu
A. nicolsonii Walker, Reed and Sanders	*E. schenckii* Sieverding and Toro
A. paulineae Blaszkowski	

Source: Compiled from http://invam.caf.wvu.edu

Appendix 7.3 AMF species of the families Archaeosporaceae(genus Archaeospora), Paraglomeraceae (genus Paraglomus) and Pacisporaceae (genus Pacispora)

Genus/*Species*
Archaeospora
Ar. trappei (Ames and Linderman) Morton and Redecker
Ar. gerdemannii (Rose, Daniels and Trappe) Morton and Redecker
Ar. leptoticha (Schenck and Smith) Morton and Redecker
Paraglomus
P. brasilianum (Spain and Miranda) Morton and Redecker
P. occultum (Walker) Morton and Redecker
Pacispora
P. boliviana Sieverd. and Oehl
P. chimonobambusae Blaszk. emend. Sieverd. and Oehl
P. coralloidea Sieverd. and Oehl
P. dominikii Sieverd. and Oehl
P. franciscana Sieverd. and Oehl
P. robigina Sieverd. and Oehl
P. scintillans Rose and Trappe emend. Sieverd. and Oehl

Source: Compiled from Oehl and Sieverding, 2004 and http://invam.caf.wvu.edu

Appendix 7.4 AMF species of the family Gigasporaceae, genera Gigaspora and Scutellospora

Genus/*Species*
Gigaspora
Gi. albida Schenck and Smith
Gi. decipiens Hall and Abbott
Gi. gigantea (Nicol. and Gerd.) Gerd. and Trappe
Gi. margarita Becker and Hall
Gi. rosea Nicol. and Schenck
Scutellospora
S. alborosea (Ferr. and Herr.) Walker and Sanders
S. arenicola Koske and Halvorson
S. armeniaca Blaszkowski
S. aurigloba (Hall) Walker and Sanders
S. biornata Spain, Sieverding and Toro
S. calospora (Nicol. and Gerd.) Walker and Sanders
S. castanea Walker
S. cerradensis Spain and Miranda
S. coralloidea (Trappe, Gerd. and Ho) Walker and Sanders
S. dipapillosa (Walker and Koske) Walker and Sanders
S. dipurpurascens Morton and Koske
S. erythropa (Koske and Walker) Walker and Sanders
S. fulgida Koske and Walker
S. gilmorei (Trappe and Herd.) Walker and Sanders
S. gregaria (Schenck and Nicol.) Walker and Sanders
S. hawaiiensis Koske and Gemma
S. heterogama (Nicol. and Gerdemann) Walker and Sanders
S. minuta (Ferr. and Herr.) Walker and Sanders
S. nigra (Redhead) Walker and Sanders
S. nodosa Blaszkowski
S. pellucida (Nicol. and Schenck) Walker and Sanders
S. persica (Koske and Walker) Walker and Sanders
S. reticulata (Koske, Miller and Walker) Walker and Sanders
S. savannicola (Ferr. and Herr.) Walker and Sanders
S. scutata Walker and Diederichs
S. spinosissima Walker, Cuenca and Sanchez
S. tricalypta (Herr. and Ferr.) Walker and Sanders
S. verrucosa (Koske and Walker) Walker and Sanders
S. weresubiae Koske and Walker

Source: Compiled from http://invam.caf.wvu.edu

Chapter 8

Saprophytic and Plant Pathogenic Soil Fungi

Ludwig H. Pfenning and Lucas Magalhães de Abreu

INTRODUCTION

Soil fungi play a key role in decomposition processes that mineralize and recycle plant nutrients. In the soil environment, fungi interact with a complex microbial community, including bacteria, actinomycetes (actinobacteria) and with small invertebrates. Fungi are also an important part of the food chain within the soil environment, mainly for the soil-inhabiting mesofauna (Bonkowski et al, 2000). In agro-ecosystems, plant pathogens act in soil and the rhizosphere causing reduction of yield and quality (Wainwright, 1988; Lodge, 1993).

To obtain a confident assessment of diversity of soil fungi there are two basic constraints. Investigations should be designed as long-term studies that engage taxonomic and other specialists in mycology, and, in addition, suitable and precise methodologies must be devised. When a long-term study involving several specialists is not feasible because of limited time and resources, the use of indicator organisms, target groups or predictor sets could be an alternative (Hyde, 1997a and b).

Agricultural activities may affect the diversity of soil-borne organisms, which play an important role in nutrient cycling or mediate the equilibrium between pathogens and their antagonists. The assessment of fungal diversity in tropical soils under different land uses is therefore one objective of the CSM-BGBD project. Broadly accepted standard methods for inventorying fungal diversity or for evaluating the impact under varying agricultural practices and other human activities are not yet available.

Classical microbiological procedures for studying soil fungi rely on culture-based procedures that involve isolation of microbial propagules or active growing hyphae from soil and growing them in axenic culture media for further identification and quantification. Methods used for the isolation of soil, rhizosphere and rhizoplane fungi have been revised by several authors (Frankland et al, 1990; Gray, 1990; Gams, 1992; Singleton et al, 1992; Cannon, 1996; Davet and Rouxel, 2000; Bills et al, 2004). A general overview on methods for studying soil-borne plant pathogenic fungi was given by Singleton et al (1992). Considerable progress has been made using washing techniques, partially selective culture media and additives that reduce the growth of certain groups of fungi.

The methodologies for isolating and culturing fungi from complex ecosystems such as soil show inherent limitations due to the fastidious nature of several species coupled with the inability of culture media to mimic soil habitats exactly (Tsao et al, 1983; Muyzer et al, 1993; Bridge and Spooner, 2001). Even the analysis of relative abundance of cultivable species recovered from soil may not adequately represent the dynamics of soil communities, since culture media impose new selective conditions and can introduce biases to the analyses (Liu et al, 1997). A reliable measure of fungal soil communities without biases requires a laborious programme of systematic isolations with different culture media and isolation strategies covering idiosyncrasies from the diverse taxonomic and physiological groups of fungi occurring in soil ecosystems. Nevertheless, generic statements that only 1 per cent of the 'micro-organisms' in soil are culturable exaggerate the difficulty and do not take into account the huge biological diversity that micro-organisms really represent (Rondon et al, 2000). Within this group we can find phylogenetically diverse groups such as the many phyla of Eubacteria, as well as quite homogeneous groups like the phylum Glomeromycota, a monophyletic group of obligate root-associated fungal organisms (see Chapter 7). However, most of the soil-inhabiting fungi must be considered saprotrophs and therefore should be able to grow in axenic culture.

Other methodologies now developed are focused on the analysis of fungal activity and its role in the biogeochemical processes occurring in soil environments. For these purposes, methods that rely on the analysis of soil microbial biomass, soil respiration, nitrogen cycling and fungal fatty acid content, or direct observations of actively growing mycelia on soil particles have all been applied (Widden and Parkinson, 1973; Houston et al, 1998; Brodie et al, 2003; Malosso et al, 2006). However these methods alone give poor information about the fungal species involved in those processes and therefore for a better understanding of soil fungal community structure and function, isolation and traditional identification procedures are still required (Brodie et al, 2003).

The results obtained from a survey of soil fungi depend largely on the methods used. In general every method gives a bias towards specific groups of fungi. Besides this we have to keep in mind that soil is not a substrate, but rather an ecosystem composed of a mixture of most diverse substrates including dead and living parts of plants, animals and other micro-organisms, along with a mineral portion and water. It is therefore all the more necessary to adopt at least similar methods within cooperative or multidisciplinary projects to guarantee future comparability of data. In this chapter we refer to some of the most commonly used methods but also present some less well established techniques for the assessment and monitoring of fungal communities in the soil environment. Principles and applications of molecular tools for soil fungal community studies are also considered. This approach is not comprehensive but aims to give a good overview on available procedures. Our main objective is to make a contribution towards the establishment of generally accepted standard methods.

Ecological importance of fungi in soil

Soil is a habitat or ecosystem rather than a substrate. This fact complicates both definitions and methodology, since soil represents a complex mixture of inorganic and organic fractions with water and living organisms. The organic fractions are composed of fresh and decaying plant material in different stages of decomposition, living roots, exudates and micro-organisms, small invertebrates and their gut contents. For this reason soil harbours a considerable part of overall fungal biodiversity and no sound estimate of numbers of soil fungal species exists (Hawksworth, 1991; Hawksworth and Rossman, 1997).

Soil fungi play a key role in decomposition processes that mineralize and recycle plant nutrients (Wainwright, 1988; Lodge, 1993; Beare et al, 1997). Saprophytes have a limited specificity for substrates; for instance, zygomycetes use simple carbohydrates, while ascomycetes principally decompose cellulose and hemicellulose (Domsch et al, 1980; Zak and Visser, 1996; Lodge, 1997). In agro-ecosystems, plant pathogens and their antagonists are particularly important. Plant pathogens act in the soil, in the rhizosphere, or infect shoots, causing damping-off in seedlings and yield losses. They can be specific, but most of them attack a wide range of host plants. There is some evidence that agricultural practices cause more quantitative than qualitative alterations in the community of soil microfungi (Pfenning, 1997; Rodrigues-Guzman, 2001). Suppressiveness of plant pathogens may be intrinsic in soils, but also can be maintained or augmented by specific agricultural practices like incorporation of organic matter, cover plants and crop diversification. Biological elements have been identified as primary factors in disease suppression (Chet and Baker, 1980; Schneider, 1984; Mazzola, 2002, 2004). It has been experimentally shown that introduction of specific antagonists like *Trichoderma* spp. or *Coniothyrium minitans* can reduce incidence of a variety of soil-borne diseases (Whipps et al, 1993).

Concerning the role of decomposer organisms, we have to mention that fungi are responsible for the degradation of xenobiotics and organic pollutants introduced in the soil (Bordjiba et al, 2001; Barratt et al, 2003; Silva et al, 2003). Fungi are also an important part of the food chain within the soil environment, mainly for the soil inhabiting mesofauna (Bonkowski et al, 2000). Maintenance of soil fungi diversity should therefore directly benefit sustainable agricultural activity, through nutrient supply, better physical structure of the soil, and antagonist control of soil-borne plant pathogens.

Aspects of modern systematics

Protozoa

Organisms called fungi are currently grouped in three kingdoms, within a phylogenetic classification system (Kirk et al, 2001). Fungal organisms in the kingdom Protozoa are not numerous, though a few of them are important soil-borne plant pathogens like *Spongospora subterranea*, *Plasmodiophora brassicae* or *Polymyxa graminis*. These genera are classified in the class Plasmodiophoromycetes. Detection of these organisms requires bio-tests or baiting. The so-called slime moulds – a heterogeneous and polyphyletic group of fungal-like organisms – live on plant debris and can also be found in soil.

Chromista

Fungal organisms which are interpreted as derived from algae and contain cellulose as the major constituent of their cell wall are classified in the kingdom Chromista or Straminipila, class Peronosporomycetes (Dick, 2001; Kirk et al, 2001). Although they represent a characteristic and phylogenetically delimited group, habitats and ecological importance are diverse. While some are true aquatic fungi associated with plant debris in water or pathogens of other aquatic organisms, others are soil inhabitants. Within the group of soil-borne Peronosporomycetes, we can find saprophytes as well as some plant pathogenic genera. Their distinct biological features require specific techniques for isolation and characterization which are presented in the section 'Culturing based procedures', below.

Glomeromycota

Fungi classified in the kingdom Fungi are characterized by having chitin as the major constituent of their cell wall. Within this kingdom, five phyla are now recognized. Genera belonging to the phylum Glomeromycota form arbuscular mycorrhizas and do not grow in the absence of their host plant (Schuessler et al, 2001; Moreira and Siqueira, 2002). Working with this group of obligate biotrophic symbionts requires different specific techniques, which are discussed in Chapter 7.

Chytridiomycota

The only group within the kingdom Fungi that forms flagellate spores is the phylum Chytridiomycota. Some of these fungi are soil inhabitants, but generally they are aquatic, living as saprophytes or parasites of other organisms like nematodes, scale insects, plants or other fungi. Several species are known as vectors of plant pathogenic viruses.

Zygomycota

Representatives of phylum Zygomycota are considered sugar-fungi because of their preference for simple carbohydrates and vigorous growth in axenic culture. Species of the genera *Absidia*, *Cunninghamella*, *Gongronella*, *Mucor*, *Mortierella* or *Rhizopus* are extremely common in soil and are reported in almost all surveys of soil fungi.

Basidiomycota

The phylum Basidiomycota is characterized by the basidium, a meiosporangium where generally four sexual spores are formed (Bauer et al, 2006), and in most cases by the formation of a fruiting body. The vegetative mycelium develops generally in soil and plant debris. Many of them form ectomycorrhizae in association with the root systems of forest trees. Assessment of diversity of these fungi as well as monitoring of impacts on the community requires specific techniques (Rossman et al, 1998) which are not considered in this review. Important soil-borne plant pathogens are species of *Rhizoctonia* and *Sclerotium*.

Ascomycota

The most numerous group of fungi is represented by the phylum Ascomycota, characterized by the ascus, a structure formed during the sexual stage of the life cycle. The asexual stage of ascomycetes is called anamorphic and is frequently the most common stage encountered. The phylogenetics of this group is still the subject of discussion and revision. For practical purposes, distinguishing major groups like plectomycetes, discomycetes, pyrenomycetes and loculoascomycetes can be useful. Some of the common soil-inhabiting ascomycetes like *Aspergillus* and *Penicillium* belong to the plectomycetes, ascomycetes which form a prototunicate ascus in a cleistothecial fruiting body. Probably the largest group of fungi which can be found in soil are pyrenomycetous anamorphs.

Soil sampling

Small amounts of soil are cored to a depth of 20cm from 12 points distributed in each sampling plot according to sampling scheme shown in Chapter 2. Each set of 12 samples is bulked to form a composite sample of about 500g and placed inside a plastic bag. Alternatively, if resources permit, three or more composite samples can be collected per sampling point. All sampling materials (corers, spatula, hoe, etc.) should be washed before and after sampling at each sampling point to avoid cross contamination. Litter must be removed just before coring takes place. For DNA extraction, soil samples should then be transferred to the laboratory in an insulated container, preferably at 4°C, and frozen as soon as possible for further processing. A second bulk sample of about 200g should be collected if other physical and chemical analysis of soil is foreseen.

Assessment of soil fungi: Culturing based procedures

All methods proposed rely on soil samples and do not contemplate the use of roots for soil-borne fungi isolation since the mycobiota of the rhizosphere is highly influenced by the plant species. The rhizosphere also contains a high concentration of bacteria, which despite the use of antibiotics in culture media, can be detrimental in the recovery of fungal species. Due to small size and few differentiated structures, detection and monitoring of fungal organisms of protists like *Spongospora* or *Plasmodiophora* must be done by bioassays using highly susceptible plants.

Peronosporomycetes (Straminipila): *Pythium* and *Phytophthora* by baiting

For recovery of these genera of Peronosporomycetes from soil the use of susceptible plants or plant tissues is indicated. Species of *Pythium* and *Phytophthora* can cause serious damage in cultivated plants. Aliquots of 2g of soil are transferred to Petri dishes containing sterilized distilled water. Baits like sorghum grains are added and plates incubated for five days. After this, baits are checked for presence of mycelium. This procedure should be continued for several weeks. Fungal mycelium formed on the bait is transferred to dishes with medium MP5 (in the case of Saprolegniaceae) and corn meal agar (CMA), potato dextrose agar (PDA) or potato carrot agar (in the case of Pythiaceae) (see Appendix 8.1 for media composition). After incubation for three to five days, pieces of the agar containing mycelium are transferred to plates containing distilled, sterilized water and two halves of sorghum seeds. At the end of the procedure it must be observed that only one isolate is obtained for each plate. The isolation of zoosporic fungi (Peronosporomycetes, Cytridiomycetes) from environmental samples is illustrated in Plate 6 in the colour section of this book. Picture 'A' shows a soil sample with baits, picture 'B' a water sample with baits, 'C' illustrates a pure culture on bait, a picture of sporangia of *Phytophthora* is seen in figure 'D', while the oogonium and antheridium of *Pythium* is depicted in picture 'E', and finally *Pythium* liberating zoospores is seen in the last picture.

Another baiting technique uses humid soil samples of 0.5g placed in sterilized glass test tubes with 3mL of sterilized water. Fragments of autoclaved grass leaf blades are added to the tubes as tissues for the recovery of *Pythium* spp. The samples are incubated for three to five days at 25°C in the dark. Infected tissues are transferred into sterile water with antibiotics (cloramphenicol) for a few hours. The growing mycelium is verified directly by water mounts in a microscope or transferred from the bait to the isolation medium CMA amended with antibiotics cloramphenicol (50mg L^{-1}) and benomyl (10mg L^{-1}) (Marks and Mitchell, 1970; Tsao et al, 1983; Gams et al, 1998; Edena et al, 2000). Quantitative assays can be made by the relative colonization frequency of the bait tissues among replicates in each soil sample. Grass leaf blades can be substituted as baits by pieces of fruits or vegetables like cucumber, tomato or potato.

Chytridiomycota by baiting

For the assessment of Chytridiomycota the same techniques are used as in Peronosporomycetes. As isolation of individual colonies in axenic culture media is difficult, characterization and identification is made directly on the bait. Most appropriate baits include shrimp shells, snake skin, insect wings and pieces of fruits and vegetables. For preservation, baits can be transferred into Castellani vials.

Basidiomycota – *Rhizoctonia*: Soil particles baiting technique

For the assessment of this genus of anamorphic basidiomycete in soil several methods are described, compiled by Sneh et al (1991). A method to evaluate inoculum density of *Rhizoctonia* spp. in eucalyptus clonal garden soils was described recently by Sanfuentes et al (2002). Two useful isolation methods are presented here.

Soil particles method

A suspension of 10g of soil sample and tap water is agitated and decanted through a 0.5mm sieve. The retained soil is re-suspended and sieved a few times further. The retained soil particles are dried on sterilized paper towel and transferred to a Petri dish of 15cm diameter containing acidified water agar medium (2 per cent) with 250mg L^{-1} of chloramphenicol. After an incubation period of 24–48 hours at 25°C typical *Rhizoctonia* growth is verified, the soil particle colonization frequency assessed and the mycelial tips transferred to PDA for an eventual characterization (Sneh et al, 1991).

Baiting technique

1g of table beet seeds is mixed with 100g of moist soil sample in Petri dishes. After an incubation period of 48 hours at 25°C, the seeds are recovered on a 1.5mm sieve, washed with distilled water for 20 minutes, dried on sterilized paper towel and transferred to water agar medium (2 per cent) containing 100mg L^{-1} of chloramphenicol. *Rhizoctonia* growth is verified and the mycelial tips transferred to PDA for isolation and characterization (Papavizas et al, 1975).

Despite the common use of several methods for isolation of *Rhizoctonia* from agricultural soil samples, their application for study of soils under natural vegetation by our research group has shown severe limitations, due to the ubiquitous presence of *Trichoderma* spp. in these soils. Thus, the employment of culture independent techniques, such as the amplification of *Rhizoctonia* ribosomal DNA with specific primers and its qualitative and even quantitative analysis can be suitable in specific circumstances (Lees et al, 2002).

Ascomycota: Soil washing and particle filter technique

This major group of soil fungi includes strict saprotrophic, entomopathogenic, plant pathogenic species and their antagonists. However, all these species exhibit a saprotrophic phase in the soil and can be isolated by a unified soil plating methodology. The soil washing technique is a suitable method for the isolation of these species.

Washing technique

A mineral soil sample of 10g is agitated with 200mL of distilled water in a shaker at 180rpm for ten minutes. After settling of soil particles for about two minutes, the supernatant suspension is discarded and the procedure repeated twice. The pre-washed soil particles are sieved in a stacked set of 1.0mm, 0.7mm, 0.5mm and 0.21mm sieves using distilled water (about two litres) for two minutes. Stable soil colloids and sand grains are then carefully taken from the last sieve, dried in sterilized paper towel and transferred (seven particles per plate), to five Petri dishes (90mm) containing the isolation medium CMA (30g L^{-1}), plus streptomycin (50mg L^{-1}) for inhibition of bacteria and cyclosporine (10mg L^{-1}) or rose Bengal (70mg L^{-1}) for inhibition of fast-growing fungi. The colonization frequency of soil particles for each fungal species is used in quantitative assays. When cyclosporine is employed an almost complete suppression of common saprotrophic zygomycetes is achieved. Therefore, when this latter group of fungi is also under study, cyclosporine usage should be avoided (Dhingra and Sinclair, 1985; Lang and Jagnow, 1986; Bååth, 1988; Gams, 1992; Bills and Polishook, 1994; Kacprzak and Stanczyk-Mazanek, 2003). This method is illustrated in Plate 7.

The soil dilution plate method

This is the most commonly used method for isolation and quantitative estimation of both bacteria and fungi. The technique is very simple and several modifications have been described. Basically, a known amount of soil is suspended in sterilized water, making a 10 per cent suspension, which is maintained under agitation for a few minutes. From this suspension a series of tenfold dilutions is prepared until the desired final dilution is achieved. A final dilution factor of 10^{-4} or 10^{-5} has been considered suitable for isolation of fungi (Dhingra and Sinclair, 1985). Aliquots of the final dilution are evenly distributed onto Petri dishes containing agar media, generally amended with antibiotics such as cloramphenicol, streptomycin or penicillin for inhibition of bacterial growth. Quantitative measurements can be achieved by multiplying the mean of colony forming units (CFU) plate count by the dilution factor employed, which gives the estimate of the number of fungal propagules per gram of soil (Bills et al, 2004). It is of general concern that the dilution plate method shows a bias towards fungi that are capable of producing large amounts of spores and grow very fast on rich culture media. Therefore the diversity of fungi that usually exist as active growing mycelia in the soil and have low ability of competing with fast-growing species when in axenic media is underestimated by this technique (Bååth, 1988; Gams et al, 1998; Tsao et al, 1983).

Ascomycota: Selective media and baiting techniques

For a selective isolation of target groups or fungal species from soil, selective media are routinely used. The selective media can contain carbon sources that are preferably metabolized by some physiological groups of target fungi or they may be amended with chemicals that inhibit the growth of undesirable organisms. A vast number of selective media have been developed for the isolation of several genera of ascomycetes, basidiomycetes and oomycetes from soil, in particular those which contain plant pathogenic species; for example, the acidified water-agar for isolation of *Rhizoctonia solani* and the Komada medium for *Fusarium* spp. (Masago et al, 1977; Tsao et al, 1983; Dhingra and Sinclair, 1985; Sneh et al, 1991; Thorn et al, 1996; Edel et al, 2001).

The selective isolation of soil fungi can also be accomplished by the use of baits that are primarily colonized by specific physiological groups of fungi. In most cases the baiting tissue is incubated with a soil sample for a few days and then transferred to a selective agar medium for the isolation of desired fungi; the colonization of the bait by a target fungus can also be accomplished by direct microscopic observations (Gams et al, 1998). Examples of baits used for selective isolation of soil fungi are plant tissues for plant pathogens, paper strips for cellulolytic species, polyester polyurethane for plastic degraders, hair pieces for keratinophilic species, chitin for chitinase producers, insect larvae for entomopathogenic and nematodes for nematophagous fungi (Marks and Mitchell, 1970; Papavizas et al, 1975; Dackman et al, 1987; Sneh et al, 1991; Gams et al, 1998; Edena et al, 2000; Gonçalves et al, 2001; Pettitt et al, 2002; Barratt et al, 2003; Wellington et al, 2003).

Baiting procedure for isolation of *Cylindrocladium* and allied genera from soil using leaves of *Ricinus communis*

Cylindrocladium is a genus that comprises saprotrophs and plant pathogenic species commonly found in soils. Despite this, species of *Cylindrocladium* and the related *Cylindrocladiella* and *Gliocladiopsis* are rarely reported from soil surveys employing the soil dilution plate method. Since *Cylindrocladium* and probably other genera are sensitive to cyclosporine they should be selectively isolated by the use of baits, and several plant materials suitable for their isolation were described in a recent monograph (Crous, 2002b). In our laboratory we have used leaves of castor oil (*Ricinus communis*) as baits for isolation of fungi from the *Cylindrocladium* complex in a routine fashion and the protocol is briefly described.

Freshly collected young leaves are washed under tap water and submitted to a superficial disinfection with 70 per cent ethanol for one minute, followed by sodium hypochloride 3 per cent for two minutes. Leaves used without previous disinfection tend to be degraded faster and colonized by bacteria. Entire leaves are put in 15cm Petri dishes and covered with a layer of the humid soil sample. After three days of incubation, the leaves are taken, washed carefully with distilled water, transferred to a humid chamber and verified daily for the typical sporulation of *Cylindrocladium*, *Cylindrocladiella*, *Gliocladiopsis* and other related genera. The isolation of fungi is accomplished by the transfer of spores to culture medium using fine glass needles under a microscope stereoscope (Gams et al, 1998; Gonçalves et al, 2001; Crous, 2002b).

Baiting technique for isolation of damping-off pathogens

A wide range of soil-borne plant pathogenic fungi can cause seedling damping-off, thus a simple baiting protocol using tomato seedlings is suitable for their isolation. Tomato seeds are germinated in 100g of soil samples and the seedlings verified for damping-off symptoms. Tissues of infected seedlings are surface sterilized with sodium hypochlorite (2 per cent for 30 seconds), washed with distilled water, dried in sterilized paper towel and transferred to isolation media such as PDA and MA (malt extract agar) containing cloramphenicol (250mg L^{-1}).

How to identify soil fungi

Identification of soil-inhabiting fungi may cause trouble to those without a specific training in mycology. We illustrate the difficulties of identifying to species level with the genus *Fusarium*, which is notorious as a heterogeneous group that includes saprophytes, endophytes and plant pathogens. The popular work of Barnett and Hunter (1998) is usable for a few of the most common saprophytic anamorphs, including soil fungi. After more than 20 years, the comprehensive work of Domsch et al (2007) is still the outstanding source for everybody who studies soil fungi. Besides the keys and monographs on about 250 genera, this compendium includes keys to species of several important genera and a compilation of literature.

Monographs and websites are available for identifying species of *Aspergillus* (Klich and Pitt 1988), *Cylindrocladium* (Crous, 2002b), *Fusarium* (Leslie et al, 2006), *Penicillium* (Pitt, 2000), *Trichoderma* (Bissett, 1984, 1991a, b, c; Samuels et al (website)), anamorphs with pigmented hyphae (Ellis, 1971, 1976) and coelomycetous anamorphs (Sutton, 1980; Nag Raj, 1993). If ascomycetes must be identified, the monographs and keys of Hanlin (1990, 1998a, b) can be a good starting point. Not specific for soil fungi, but nevertheless useful for this topic are some other compilations and monographs like Ellis and Ellis (1985), Samson et al (2000) or the coursebook of Gams et al (1998). Monographs are available for chytrids (Karling 1977) as well as for the genera *Pythium* (Waterhouse, 1967, 1968; van der Plaats-Niterink, 1981) and *Phytophthora* (Waterhouse, 1970; Tsao et al, 1983; Erwin and Ribeiro, 1996). The monograph of Sneh et al (1991) is useful for work with *Rhizoctonia*. For further instructions on how to identify fungi of more specific groups the *Dictionary of the Fungi* (Kirk et al, 2001) must be consulted. A separate list of references of handbooks and manuals for identification of particular groups of soil fungi is given at the end of this chapter.

ASSESSMENT OF SOIL FUNGI: DNA TARGETED TECHNIQUES

A reliable measure of soil fungal communities requires a laborious programme of systematic isolations with different culture media and isolation strategies reflecting the very diverse taxonomic and physiological groups of fungi occurring in soil (Tsao et al, 1983; Muyzer et al, 1993; Bridge and Spooner, 2001; Bills et al, 2004). However, some fungal groups such as Basidiomycota are difficult to isolate and may fail to sporulate when in axenic media (Thorn et al, 1996). Other methodologies can be used to complement the traditional approaches for a better understanding of diversity and dynamics of soil fungi and these include DNA target techniques. Molecular tools based on DNA analysis have been successfully applied for the study of complex bacterial and, recently, fungal assemblages from environmental samples. DNA sequences are directly amplified from soil by the polymerase chain reaction (PCR), and further characterized by several approaches (Bridge and Spooner, 2001).

DNA extraction

A crucial step prior to the PCR is DNA extraction directly from soil. The fungal cells present in the soil sample, even as mycelia or spores, have to be correctly lysed. The extraction protocols are mainly based on the physical process of bead beating associated with heat and chemical lyses. The crude total DNA thus obtained must be thoroughly purified to eliminate humic and phenolic substances that interfere with the PCR. This step is achieved by column separations or the use of commercial DNA purification kits (Liu et al, 1997; Viaud et al, 2000; Bridge and Spooner, 2001). Kits for extraction of total DNA from soil are commercially available. The micro-scale heterogeneities found in soil must be considered when choosing the size of soil sample submitted to total DNA extraction. Very small samples (less than 1g) should be used with caution since these are more error prone due to higher within-sample variability (Ranjard et al, 2001).

Polymerase chain reaction 'fungal' primers

The gene cluster for ribosomal RNA molecules 18S, 5.8S and 28S is thought to be highly conserved among eukaryotic organisms and is usually employed as molecular marker. DNA length polymorphisms and base sequence variations can be used to group organisms according to their origin and evolutionary relationship. The identification of unknown DNA sequences can also be accomplished by comparing them with nucleotide sequence databases from putatively known taxa. Several copy numbers of the rDNA clusters are found in the eukaryotic genome, facilitating the amplification from very small DNA samples. The rDNA cluster also contains spacers between the coding regions known as the internal transcribed spacers (ITS) which have less conserved base sequences and can be used for differentiation among related species or to assess infraspecific genetic divergences (Viaud et al, 2000; Baayen et al, 2000; Down, 2002).

Since the soil contains a vast number of organisms, obtaining specific fungal rDNA primers is a critical step in PCR amplification. The primers must permit amplification of a broad range of fungal species without losing the specificity to this target group. Based on the ribosomal RNA gene sequences available on specialized databases for several fungal species, primers that are specific for PCR amplification of fungal DNA from complex soil samples have been developed and were compiled by Anderson and Cairney (2004). However, some of these specific primers have been shown to amplify non-fungal DNA or show bias towards the amplification of particular taxonomic groups inside the kingdom Fungi (Smit et al, 1999; Borneman and Hartin, 2000; Anderson et al, 2003).

Detection and/or quantification of a specific few fungal species is generally required in the field of plant pathology. For this type of research several species-specific primers have been developed for direct detection and monitoring of target plant pathogens from soil samples (Cullen et al, 2002; Lees et al, 2002; Filion et al, 2003). Several fungi present in very low densities in natural soils, such as *Phytophthora* species, are difficult to detect by molecular means since PCR tends to amplify DNA molecules that are dominant in the total extract DNA. For these fungi, molecular detection from soil samples can be improved through baiting techniques coupled with PCR using genus or species-specific primers (Nechwatal et al, 2001).

DNA sequencing

In order to obtain the identities of the PCR-amplified fungal DNA, the amplicons of correct size can be separated in agarose gels, excised from the gel matrix, purified, connected to plasmid vectors and cloned into bacterial cells. The cloned DNA is sequenced and compared with databases containing fungal rDNA oligonucleotide sequences via software analyses. These procedures are expensive, time-consuming and not suitable for the assessment of complex environmental samples. Therefore, molecular community fingerprint techniques have been developed (Borneman and Hartin, 2000; Viaud et al, 2000; Vandenkoornhuyse et al, 2002; Anderson et al, 2003).

Molecular fingerprinting

TGGE and DGGE

The thermal or denaturing gradient gel electrophoresis was introduced in studies on microbial ecology by Muyzer and Smalla (1998). In this technique, PCR-amplified 18S DNA molecules are submitted to an electrophoresis in a gel matrix with a linear thermal (TGGE) or denaturing (DGGE) gradient. Amplicons have exactly the same number of nucleotides, but differ in *guanine-cytosine* content (GC), which determines melting behaviour and electrophoretic mobility when migrating in the denaturating gel. The primers used in the PCR generally contain a DNA clamp built up of repeated GC bases which are less prone to denaturing due to their higher chemical stability. The denaturing gradient in DGGE can be varied by the use of different amounts of urea and formamide in the gel matrix. The band patterns generated are used to analyse complex soil fungal communities. Under these denaturing conditions, DNA stretches of the same length (the common output of a PCR) but with different base-pair sequences, partially melted due to the differential denaturing of less stable domains in the molecules, are called melting domains. Partially melted molecules halt their movement in the gel. Therefore, DNA molecules with different base-pair sequences have different melting behaviour and migrate to distinct points in the gel (Muyzer et al, 1993). Several studies used this technique to assess community structures of fungi and other micro-organisms and monitor modifications or impact due to agricultural practices or industrial activities (Smit et al, 1999; Elsas et al, 2000; Gomes et al, 2003; Milling et al, 2004). Constraints and limitations are represented by the choice of primers used in the first amplification step after the extraction of total DNA from an environmental sample. Several primer combinations have already been proposed for amplification of DNA from distinct groups of fungi like Ascomycota, Basidiomycota, Chytridiomycota, Zygomycota and Oomycota (Nikolcheva and Bärlocher, 2004). The use of a nested PCR can give a better result still and increase resolution in separation of the PCR products (Oros-Sichler et al, 2006). Another constraint is that DNA from phylogenetically distant organisms can produce PCR products with identical electrophoretic mobility (Gomes et al, 2003). Nevertheless, the technique can provide a highly reproducible standard from a large number of environmental samples in a relatively short time. More recently, polyphasic approaches that include construction of cloned fragment libraries, molecular fingerprinting, biochemical analyses and isolation techniques are becoming the procedures of choice (Malosso et al, 2006; Singh et al, 2006).

PCR-RFLP

The low variability of fungal rDNA may introduce artefacts in molecular fingerprint techniques such as TGGE and DGGE, for example co-migration of different DNA molecules to the same point in the gel. To increase the band polymorphism and improve the resolution, enzymatic restriction of DNA prior to electrophoresis can be used in a technique known as restriction fragment length polymorphism (RFLP) of PCR-amplified DNA (Viaud et al, 2000). Related fungal species can be identified by their typical RFLP patterns after PCR amplification of DNA directly extracted from soil using genus-specific primers (Nechwatal et al, 2001).

T-RFLP

Liu et al (1997) developed an extended version of the PCR-RFLP, called terminal restriction fragment length polymorphism, for the analysis of bacterial communities from environmental samples. In the T-RFLP, the DNA is PCR-amplified with one of the two primers labelled with fluorescence. After amplification, the DNA is digested with restriction enzymes and the restriction fragments separated in a gel according to their sizes. The terminal restriction fragments (TRFs) are labelled with fluorescence and can be detected and quantified in an automated sequencer. A database of TRFs from known fungal species can be constructed and used for taxonomic correlation with the haplotypes detected in the T-RFLP analysis (Brodie et al, 2003; Edel-Hermann et al, 2004; Kennedy et al, 2005; Singh et al, 2006)

ARISA

The natural polymorphism in the ITS region of the rDNA cluster among fungal species can be used for the evaluation of soil fungal communities. In the automated ribosomal intergenic spacer analysis (ARISA), the fungal DNA is extracted from soil and PCR-amplified using specific primers for the ITS region with one of the primers labelled with fluorescence. Following the amplification, the different ITS stretches are separated in a gel, detected and quantified by an automated sequencer (Ranjard et al, 2001). The fungal ARISA is a highly sensitive technique; however, intra-specific ITS polymorphism may present problems in the separation between distinct phylogenetic species (O'Donnell and Cigelnik, 1997; Viaud et al, 2000; Kennedy et al, 2005).

SSCP

Another molecular fingerprint approach uses the differential migration of single-stranded DNA molecules in a gel to study complex microbial communities. In single strand conformation polymorphism (SSCP), the amplified DNA is fully denatured before being submitted to an electrophoresis run. Single-stranded DNA molecules acquire unique folded structures dictated by their nucleotide sequences and migrate to specific points in the gel. Consequently, DNA fragments with the same length but dissimilar in base-pair sequence can be readily separated in SSCP by their differential migration in the gel when in single-strand conformation (Lee et al, 1996; Lowell and Klein 2001; He et al, 2005). Table 8.1 shows some examples of molecular techniques employed for the study of soil fungal communities published in the recent years.

Quantitative PCR

Some studies, especially with soil-borne plant pathogenic fungi, are focused on only one or a few fungal species present in the soil. For such studies, very specific PCR primers must be designed for correct amplification of DNA from the target species, and other molecular markers can be employed such as the sequences of β-tubulin gene or the translation elongation factor EF 1-α gene (Baayen et al, 2000; Mauchline et al, 2002; Filion et al, 2003; Li and Hartman 2003). Besides molecular detection, the quantification of fungal DNA in soil is generally required in epidemiological and ecological studies. However, the conventional PCR is not suitable for quantitative approaches since small variations during the exponential phase of the amplification reaction can drastically alter the amounts of PCR products. For quantitative assays, modifications of conventional PCR as competitive PCR (cPCR) and real time PCR have been developed.

In cPCR, DNA fragments containing the same primer sites as the sample DNA are added in known amounts to the PCR reactions and co-amplified with the target DNA. After amplification the distinct PCR products are separately quantified by their relative band intensities in agarose gels. By the addition of different amounts of competitive DNA to a standard amount of sample DNA in a series of PCRs, and with the monitoring of competitive PCR products yielded in each reaction, the amount of DNA present in the sample can be calculated (Siebert and Larrick, 1992; Baek and Kenerley, 1998; Mauchline et al, 2002).

Another method employed in quantitative approaches is the real time PCR. Fluorogenic probes or dyes added to the PCRs permit the correct monitoring of products yielded during the amplification. The added fluorescent markers are capable of emitting fluorescence in the presence of double-stranded DNA; therefore the increase of DNA molecules during the amplification leads to an increase in the intensity of fluorescent emission, which can be calibrated and accurately measured. A standard curve can be constructed with known amounts of DNA and the fluorescent intensities after a defined number of PCR cycles, and used to quantify DNA directly from soil samples. This technique is faster and simpler than cPCR since there is no need for the construction of competitor DNA and no post-PCR analysis is required (Cullen et al, 2002; Lees et al, 2002; Filion et al, 2003).

Table 8.1 *Compilation of approaches using DNA cloning and sequencing for the identification of the dominant haplotypes, amplified from total DNA*

*Technique**	*Assessment of fungal communities in*	*References*
TGGE	Rhizosphere and bulk soil under wheat	Smit et al, 1999
PCR-RFLP	A soil sample by culture-based and molecular methods	Viaud et al, 2000
SSCP	Short-grass steppe soil with or without nitrogen amendment	Lowell and Klein, 2001
ARISA	Soils with different physicochemical properties	Ranjard et al, 2001
DGGE	Analysis of the dynamics of fungal communities in soil via fungal-specific PCR of soil DNA followed by DGGE	Elsas et al, 2000
DGGE	A gradient of soil from moorland to Scots pine forest	Anderson et al, 2003
DGGE and TRFLP	Gradient from semi-natural grassland to agriculturally improved soils	Brodie et al, 2003
DGGE and restriction of amplified DNA	Rhizosphere of two maize cultivars during the period of plants' life cycle	Gomes et al, 2003
DGGE	The different horizons of a forest soil	Agnelli et al, 2004
T-RFLP	Soils with different physicochemical properties; effects of soil amendment with compost or manure	Edel-Hermann et al, 2004
DGGE	Soils under transgenic and non-transgenic potato cultivars	Milling et al, 2004
SSCP and TGGE	Natural forest and hoop pine plantation soils	He et al, 2005
ARISA and T-RFLP	Grasslands under different vegetations	Kennedy et al, 2005
DGGE with primers specific for *Fusarium*	Asparagus fields; specific detection and diversity analysis of *Fusarium* species	Yergeau et al, 2005
DGGE and T-RFLP	Three different soils under *Lolium perenne*	Singh et al, 2006
DGGE and ARDRA	Fungal diversity in Antarctic soils combining culture isolation, molecular fingerprinting and cloning techniques	Malosso et al, 2006

Note: * DNA cloning and sequencing was employed by all authors for the identification of the dominant haplotypes amplified from total DNA, thus this method is not described for each example.

Conclusions

Molecular tools for soil fungal community assays were derived mainly from studies on bacterial diversity and generally presuppose the same limitations of culture-based techniques for fungi (Selenska and Klingmüller, 1992; Tsai and Olson, 1992; O'Donnell et al, 1994). However, in the case of fungi the number of non-cultivable species is surely overemphasized. What can be true for strictly symbiotic organisms like arbuscular mycorrhizal fungi, does not apply for saprotrophic fungi. Even most of the known plant pathogenic fungi are not obligate parasites. Many constraints to culture-based identification have been overcome with the advent of more sophisticated isolation techniques like soil washing and particle filtration methods, as already discussed. Community fingerprint techniques can monitor complex fungal assemblages from soil, but a better characterization and phylogenetic analysis of dominant taxa invariably requires sequencing and comparison with public DNA databanks. However, fungal taxonomic approaches based on molecular data only may present some problems due to the incompleteness of databanks and the presence of misidentified DNA sequences deposited, originating from incorrect morphological identifications, amplification and sequencing of fungal contaminants or even sequencing of chimeras (Crous, 2002a; Bridge et al, 2003; Hawksworth, 2004b).

Another limitation of molecular approaches based on DNA amplification is the lack of knowledge concerning actively growing or particular functional groups of soil fungi in the total DNA extracted from soil. As already discussed, the DNA extraction protocols are capable of lysing and extracting nucleic acids either from actively growing mycelia or dormant spores, making the differentiation of active growth or functional groups based only on PCR product analyses difficult. For now, a more complete understanding of fungal communities in soil should rely on polyphasic approaches coupling culture-based and molecular taxonomic assays in association with the study of major biogeochemical processes occurring in soil (Malosso et al, 2006; Singh et al, 2006).

Data analysis

A list of species or species aggregates isolated and their relative abundance in each sample generally constitutes the first raw dataset obtained. The number of isolates from each sample is ascertained and can be used for comparisons between sites, collection dates or different soil treatments. The chi squared test of independence can be used for this analysis, where the total number of isolates of each sample is used in pair-wise comparisons. Where data on quantitative variations among repetitions within the samples are available, a more reliable approach is to perform an analysis of variance (ANOVA). In ANOVA it is possible to derive variations between different samples and within samples (error); thus it is a robust and confident analysis (Sokal and Rolf, 1995). In addition, other factors beside the number of isolates, such as physical and chemical parameters of the soils under analysis, can be added to the ANOVA for evaluation of their sole effects or their interactions with other quantitative data (Setälä and McLean, 2004). The raw dataset is generally not suitable for direct analysis using ANOVA, since some of the data may not be normally distributed. To circumvent this problem, data transformations such as log or square root transformations can be applied before the analysis (Houston et al, 1998). Another approach is to perform a non-parametric analysis of variance such as Kruskall–Wallis (Rodrigues, 1994; Sokal and Rohlf, 1995).

Simple species accumulation or rarefaction curves can provide graphical comparisons of species richness among samples, and also give insights about the effectiveness of the sampling strategy employed in the research. The slope of the curves shows which samples are more species rich, for example those curves that build up more readily as the number of sampled isolates accumulates. The shape of curves shows if the number of species isolated is stable after accumulation of all isolates, for example if the curves tend to reach an asymptote. Curves that do not tend to an asymptote indicate that the number of species isolated was more likely to increase if more sampling effort was made (Bills and Polishook, 1994).

Species diversity of any given system is made of two components, the species richness (e.g. the number of species) and the evenness or equitability of species occurrence (Kennedy and Smith, 1995). Many indices have been derived to evaluate and compare species diversity, based on species richness, evenness or taking both variables into account, and they vary in the relative importance given to dominant, common or rare species and complexity of the mathematical models assumed to underlie species distributions. The Shannon–Wiener index is based on richness and equitability of species diversity and is commonly found in the literature (Dighton, 1994; Persiani et al, 1998). Similarity (dissimilarity) indices of diversity for direct pair-wise comparisons among different samples are also commonly used (Magurran, 1988; Bettucci et al 1993; Grishkan et al, 2003). Diversity indices are purely numerical and give no information about community structure, and cases occur where two different communities can have the same diversity index. Community structure information is best devised by the use of multivariate analyses such as ordination methods. Correspondence analysis (CA), and its modified version, detrended correspondence analysis, are ordination methods suitable for the kinds of data generated from fungal or other communities, for example data generated from counts, large data matrices 'full of zeros' and non-parametric distribution of data. In CA, as well in other ordination techniques, the multidimensional relationships among samples and species and between samples and species are reduced to a few component axes that, ideally, explain most data variability. The results are displayed graphically allowing the analysis of general trends of community structure (Gauch, 1982; Howard and Robinson, 1995). The percentage of total variance explained by the first axes of CA and the contribution of species and samples to the variance (inertia) of each axis gives support to the designation of species occurring in the community as key and 'casual' (Howard and Robinson, 1995). Generally, the multivariate analysis is not performed on the whole community due to the 'noisy' nature of the occurrence and distribution of rare species. Thus a limit is chosen (e.g. 0.5 per cent, 1 per cent or 5 per cent of total number of isolates), and only those species with frequency of isolation equal or superior to this limit are included in the analysis (Gauch, 1982; Bettucci and Alonso, 1995; Bettucci et al, 2002). The ordination methods of principal component analysis and correspondence analysis as well as cluster analyses are usually employed for the results derived from molecular fingerprint methods, where matrices are constructed with data on the presence or absence of distinct bands (ideally different species) in the gels and the community structure is further analysed (Fromin et al, 2002; Oros-Sichler et al, 2006).

Preservation and genetic resource collection

Inventories of biodiversity furnish arguments and some scientific basis for the preservation of habitats and sustainable land use and therefore need to be compiled. Genetic resource collections are now requested in all parts of the world with the aim of *ex situ* preservation of species and to supply research institutions and industry with authentic material. Reference collections must be supported by countries with a strong policy in science and technology because they contain information on geographical and host distribution and provide basic working material for those studying characteristics and variation, as well as practical and economic applications of species (Smith and Waller, 1992; Hawksworth, 1993; Kirsop, 1996). The current status of fungal genetic resource collections, and the challenges to support the needs of fungal genomics, molecular biology and conservation were reviewed recently (Hawksworth, 1996, 2004a; Ryan and

Smith, 2004). In the future, long-term preservation of voucher specimens and cultures will be essential to validate entries in sequence and genomic databases, as is already the case for publishing new names (Agerer et al, 2003).

Acknowledgements

We are grateful to Professor Richard Mibey and Dr Sheila Okoth for comments and discussions during the preparation of the chapter.

References

Agerer, R., Ammirati, J., Blanz, P., Courtecuisse, R., Dejardin, D. E., Gams, W., Hallenberg, N., Halling, R., Hawksworth, D. L., Horak, E., Korf, R. P., Mueller, G. M., Oberwinkler, F., Rambold, G., Summerbell, R. C., Triebel, D. and Watling, R. (2003) 'Always deposit vouchers', *Mycological Research*, vol 104, pp642–644

Agnelli, A., Ascher, J., Cortia, G., Ceccherini, M. T., Nannipieri, P. and Pietramellar, G. (2004) 'Distribution of microbial communities in a forest soil profile investigated by microbial biomass, soil respiration', *Soil Biology and Biochemistry*, vol 36, pp859–868

Anderson, I. C. and Cairney, J. W. G. (2004) 'Diversity and ecology of soil fungal communities: Increased understanding through the application of molecular techniques', *Environmental Microbiology*, vol 6, pp769–779

Anderson, I. C., Campbell, C. D. and Prosser, J. I. (2003) 'Potential bias of fungal 18S rDNA and internal transcribed spacer polymerase chain reaction primers for estimating fungal biodiversity in soil', *Environmental Microbiology*, vol 5, pp36–47

Bååth, E. (1988) 'A critical examination of the soil washing technique with special reference to the effect of the size of the soil particles', *Canadian Journal of Botany*, vol 66, pp1566–1569

Baayen, R. P., O'Donnell, K., Bonants, P. J. M., Cigelnik, E., Kroon, L. P. N., Roebroeck, E. J. A. and Waalwijk, C. (2000) 'Gene genealogies and AFLP analyses in the *Fusarium oxysporum* complex identify monophyletic and nonmonophyletic formae speciales causing wilt and rot disease', *Phytopathology*, vol 90, pp891–900

Baek, J. and Kenerley, C. M. (1998) 'Detection and enumeration of a genetically modified fungus in soil environments by quantitative competitive polymerase chain reaction', *FEMS Microbiology Ecology*, vol 25, pp419–428

Barratt, S. R., Ennos, A. R., Greenhalgh, M., Robson, G. D. and Handley, P. S. (2003) 'Fungi are the predominant micro-organisms responsible for degradation of soil-buried polyester polyurethane over a range of soil water holding capacities', *Journal of Applied Microbiology*, vol 95, pp78–85

Bauer, R., Begerow, D., Sampaio, J. P., Weiss, M. and Oberwinkler, F. (2006) 'The simple-septate basidiomycetes: A synopsis', *Mycological Progress*, vol 5, pp41–66

Beare, M. H., Vikram Reddy, M., Tian, G. and Srivastava, S. C. (1997) 'Agricultural intensification, soil biodiversity and agroecosystem function in the tropics: the role of decomposer organisms', *Applied Soil Ecology*, vol 6, pp87–108

Bettucci, L. and Alonso, R. (1995) 'The effect of wildfire on the opportunistic decomposer fungal community of Uruguayan *Eucalyptus* spp. forest', *Pedobiologia*, vol 39, pp470–480

Bettucci, L., Malvarez, I., Dupont, J., Bury, E. and Roquebert, M. F. (2002) 'Paraná river delta wetlands soil microfungi', *Pedobiologia*, vol 46, pp606–623

Bettucci, L., Rodriguez, C. and Indarte, R. (1993) 'Fungal communities of two grazing-land soils in Uruguay', *Pedobiologia*, vol 37, pp72–82

Bills, G. F. and Polishook, J. D. (1994) 'Abundance and diversity of microfungi in leaf litter of a lowland rain forest in Costa Rica', *Mycologia*, vol 86, pp187–198

Bills, G. F., Christensen, M., Powell, M. and Thorn, G. (2004) 'Saprobic soil fungi', in G. M. Mueller, G. F. Bills and M. S. Foster (eds) *Biodiversity of Fungi: Inventory and Monitoring Methods*, Elsevier, Amsterdam, pp271–302

Bonkowski, M., Griffith, B. S. and Ritz, K. (2000) 'Food preference of earthworms for soil fungi', *Pedobiologia*, vol 44, pp666–676

Bordjiba, O., Steiman, R., Kadri, M., Semadi, A. and Guiraud, P. (2001) 'Removal of herbicides from liquid media by fungi isolated from a contaminated soil', *Journal of Environmental Quality*, vol 30, pp418–426

Borneman, J. and Hartin, R. J. (2000) 'PCR primers that amplify fungal rRNA genes from environmental samples', *Applied and Environmental Microbiology*, vol 66, pp4356–4360

Bridge, P. D. and Spooner, B. M. (2001) 'Soil fungi, diversity and detection', *Plant and Soil*, vol 232, pp147–154

Bridge, P. D., Roberts, P. J., Spooner, B. M. and Panchal, G. (2003) 'On the unreliability of published DNA sequences', *New Phytologist*, vol 160, pp43–48

Brodie, E., Edwards, S. and Clipson, N. (2003) 'Soil fungal community structure in a temperate upland grassland soil', *FEMS Microbiology Ecology*, vol 45, pp105–114

Cannon, P. F. (1996) 'Filamentous fungi', in G. Hall (ed) *Methods for the Examination of Organismal Diversity in Soils and Sediments,* CAB International, Wallingford, pp127–145

Chet, I. and Baker, K. F. (1980) 'Induction of suppressiveness to *Rhizoctonia solani* in soil', *Phytopathology*, vol 70, pp994–998

Crous, P. W. (2002a) 'Adhering to good cultural practice', *Mycological Research*, vol 106, pp1378–1379

Crous, P. W. (2002b) *Taxonomy and Pathology of Cylindrocladium (Calonectria) and Allied Genera*, APS Press, St Paul, MN

Cullen, D. W., Lees, A. K., Toth, I. K. and Duncan, J. M. (2002) 'Detection of *Colletotrichum coccodes* from soil and potato tubers by conventional and quantitative real-time PCR', *Plant Pathology*, vol 51, pp281–292

Dackman, C., Olsson, S., Jansson, H., Lundgren, B. and Nordbring-Hertz, B. (1987) 'Quantification of predatory and endoparasitic nematophagous fungi in soil', *Microbial Ecology*, vol 13, pp89–93

Davet, P. and Rouxel, F. (2000) *Detection and Isolation of Soil Fungi*, Science Publishers, Enfield

Dhingra, O. D. and Sinclair, J. B. (1985) *Basic Plant Pathology Methods*, CRC Press Inc., Boca Raton

Dick, M. W. (2001) *Straminipilous Fungi*, Kluwer Academic Publishers, Dordrecht, The Netherlands

Dighton, J. (1994) 'Analysis of micromycete communities in soil: A critique of methods', *Mycological Research*, vol 98, pp796–798

Domsch, K. H., Gams, W. and Anderson, T. (2007) *Compendium of Soil Fungi, 2nd edition*, IHW Verlag, Eching

Down, G. (2002) 'Fungal family trees: Finding relationships from molecular data', *Mycologist*, vol 16, pp51–58

Edel, V., Steinberg, C., Gautheron, N., Recorbet, G. and Alabouvette, G. (2001) 'Genetic diversity of *Fusarium oxysporum* populations isolated from different soils in France', *FEMS Microbiology Ecology*, vol 36, pp61–71

Edel-Hermann, V., Dreumont, C., Pérez-Piqueres, A. and Steinberg, C. (2004) 'Terminal restriction fragment length polymorphism analysis of ribosomal RNA genes to assess changes in fungal community structure in soils', *FEMS Microbiology Ecology*, vol 47, pp397– 404

Edena, M. A., Hillb, R. A. and Galpoththage, M. (2000) 'An efficient baiting assay for quantification of *Phytophthora cinnamomi* in soil', *Plant Pathology*, vol 49, pp515–522

Elsas, J. D., Duarte, G. F., Keijzer-Wolters, A. and Smit, E. (2000) 'Analysis of the dynamics of fungal communities in soil via fungal-specific PCR of soil DNA followed by denaturing gradient gel electrophoresis', *Journal of Microbiological Methods*, vol 43, pp133–151

Filion, M., St-Arnaud, M. and Jabaji-Hare, S. H. (2003) 'Direct quantification of fungal DNA from soil substrate using real-time PCR', *Journal of Microbiological Methods*, vol 53, pp67–76

Frankland, J., Dighton, J. and Boddy, L. (1990) 'Methods for studying fungi in soil and forest litter', in R. G. Grigorova and J. R. Norris (eds) *Methods in Microbiology*, vol 22, Academic Press, London, pp343–404

Fromin, N., Hamelin, J., Tarnawski, S., Roesti, D., Jourdain-Miserez, K., Forestier, N., Teyssier-Cuvelle, S., Gillet, F., Aragno, M. and Rossi, M. (2002) 'Statistical analysis of denaturing gel electrophoresis (DGE) fingerprinting patterns', *Environmental Microbiology*, vol 11, pp634–643

Gams, W. (1992) 'The analysis of communities of saprophytic microfungi with special reference to soil fungi', in W. Winterhoff (ed) *Fungi in Vegetation Science*, Kluwer Academic Publishers, Dordrecht, pp183–223

Gams, W., Hoekstra, E. S. and Aptroot, A. (1998) *CBS Course of Mycology*, 4th edition, CBS, Baarn, The Netherlands

Gauch, H. G., Jr (1982) *Multivariate Analysis in Community Ecology*, Cambridge University Press, New York

Gomes, N. C. M., Fagbola, O., Costa, R., Rumjanek, N. G., Buchner, A., Mendonça-Hagler, L. and Smalla, K. (2003) Dynamics of fungal communities in bulk and maize rhizosphere soil in the tropics, *Applied and Environmental Microbiology*, vol 69, pp3758–3766

Gonçalves, R. C., Alfenas, A. C., Maffia, L. A. and Crous, P. W. (2001) 'Evaluation of bioassays to quantify *Cylindrocladium* inocula soil', *Mycoscience*, vol 41, pp261–264

Gray, T. R. G. (1990) 'Methods for studying the microbial ecology of soil', in R. G. Grigorova and J. R. Norris (eds) *Methods in Microbiology*, vol 22, Academic Press, London, pp310–342

Grishkan, I., Nevo, E., Wasser, S. P. and Beharav, A. (2003) 'Adaptive spatiotemporal distribution of soil microfungi in 'Evolution Canyon' II, Lower Nahal Keziv, western Upper Galilee, Israel', *Biological Journal of the Linnean Society*, vol 78, pp527–539

Hawksworth, D. L. (1991) 'The fungal dimension of biodiversity, magnitude, significance and conservation', *Mycological Research*, vol 95, pp641–655

Hawksworth, D. L. (1993) 'The tropical fungal biota, census, pertinence, prophylaxis, and prognosis', in S. Isaac, J. C. Frankland, R. Watling and A. J. S. Whalley (eds) *Aspects of Tropical Mycology*, Cambridge University Press, Cambridge

Hawksworth, D. L. (1996) 'Microbial collections as a tool in biodiversity and biosystematic research', in R. A. Samson, J. A. Stalpers, D. van der Mei and A. H. Stouthamer (eds) *Culture Collections to Improve Quality of Life*, Proceedings of the 8th International Congress for Culture Collections, CBS, Baarn, pp26–35

Hawksworth, D. L. (2004a) 'Fungal diversity and its implications for genetic resource collections', *Studies in Mycology*, vol 50, pp9–18

Hawksworth, D. L. (2004b) '"Misidentifications" in fungal DNA sequence databanks', *New Phytologist*, vol 161, pp13–15

Hawksworth, D. L. and Rossman, A. Y. (1997) 'Where are all the undescribed fungi?', *Phytopathology*, vol 87, pp888–891

He, J., Xu, Z. and Hughes, J. (2005) 'Analyses of soil fungal communities in adjacent natural forest and hoop pine plantation ecosystems of subtropical Australia using molecular approaches based on 18S rRNA genes', *FEMS Microbiology Letters*, vol 247, pp91–100

Houston, A. P. C., Visser, S. and Lautenschlager, R. A. (1998) 'Microbial processes and fungal community structure in soils from clear-cut and unharvested areas of two mixed wood forests', *Canadian Journal of Botany*, vol 76, pp630–640

Howard, P. J. A. and Robinson, C. H. (1995) 'The use of correspondence analysis in studies of successions of soil organisms', *Pedobiologia*, vol 39, pp518–527

Hyde, K. D. (ed) (1997a) *Biodiversity of Tropical Microfungi*, Hong Kong University Press, Hong Kong

Hyde, K. D. (1997b) 'Can we rapidly measure fungal diversity?', *Mycologist*, vol 11, pp176–178

Kacprzak, M. and Stanczyk-Mazanek, E. (2003) 'Changes in the structure of fungal communities of soil treated with sewage sludge', *Biology and Fertility of Soils*, vol 38, pp89–95

Kennedy, A. C. and Smith, K. L. (1995) 'Soil microbial diversity and the sustainability of agricultural soils', *Plant and Soil*, vol 170, pp75–86

Kennedy, N., Edwards, S. and Clipson, N. (2005) 'Soil bacterial and fungal community structure across a range of unimproved and semi-improved upland grasslands', *Microbial Ecology*, vol 50, pp463–473

Kirk, P. M., Cannon, P. F., David, J. C. and Stalpers, J. A. (2001) *Dictionary of the Fungi*, 9th edition, CAB International, Wallingford

Kirsop, B. (ed) (1996) *Access to Ex-situ Microbial Genetic Resources within the Framework to the Convention on Biological Diversity*, World Federation for Culture Collections, WFCC

Lang, E. and Jagnow, G. (1986) 'Fungi of a forest soil nitrifying at low pH values', *FEMS Microbiology Ecology*, vol 38, pp257–265

Lee, D., Zo, Y. and Kim, S. (1996) 'Nonradioactive method to study genetic profiles of natural bacterial communities by PCR single-strand-conformation polymorphism', *Applied and Environmental Microbiology*, vol 62, pp3112 –3120

Lees, A. K., Cullen, D. W., Sullivan, L. and Nicholson, M. J. (2002) 'Development of conventional and quantitative real-time PCR assays for the detection and identification of *Rhizoctonia solani* AG-3 in potato and soil', *Plant Pathology*, vol 51, pp293–302

Li, S. and Hartman, G. L. (2003) 'Molecular detection of *Fusarium solani* f. sp. *glycines* in soybean roots and soil', *Plant Pathology*, vol 52, pp74–83

Liu, W., Marsh, T. L., Cheng, H. and Forney, L. (1997) 'Characterization of microbial diversity by determining terminal restriction fragment length polymorphisms of genes encoding 16S rRNA', *Applied and Environmental Microbiology*, vol 63, pp4516–4522

Lodge, D. J. (1993) 'Nutrient cycling by fungi in wet tropical forests', in S. Isaac, J. C. Frankland, R. Watling and A. J. S. Whalley (eds) *Aspects of Tropical Mycology*, Cambridge University Press, Cambridge, pp37–58

Lodge, D. J. (1997) 'Factors related to diversity of decomposer fungi in tropical forests', *Biodiversity and Conservation*, vol 6, pp681–688

Lowell, J. L. and Klein, D. A. (2001) 'Comparative single-strand conformation polymorphism (SSCP) and microscopy-based analysis of nitrogen cultivation interactive effects on the fungal community of a semiarid steppe soil', *FEMS Microbiology Ecology*, vol 36, pp85–92

Magurran, A. E. (1988) *Ecological Diversity and its Measurement*, Princeton University Press, Princeton

Malosso, E., Waite, I. S., English, L., Hopkins, D. W. and O'Donnell, A. G. (2006) 'Fungal diversity in maritime Antarctic soils determined using a combination of culture isolation, molecular fingerprinting and cloning techniques', *Polar Biology*, vol 29, pp552–561

Marks, G. C. and Mitchell, J. E. (1970) 'Detection, isolation and pathogenicity of *Phytophthora megasperma* from soils and estimation of inoculum levels', *Phytopathology*, vol 60, pp1687–1690

Masago, H., Yoshikawa, M., Fukada, M. and Nakanishi, N. (1977) 'Selective inhibition of *Pythium* spp. on a medium for direct isolation of *Phytophthora* spp. from soils and plants', *Phytopathology*, vol 67, pp425–428

Mauchline, T. H., Kerry, B. R. and Hirsch, P. R. (2002) 'Quantification in soil and the rhizosphere of the nematophagous fungus *Verticillium chlamydosporium* by competitive PCR and comparison with selective plating', *Applied and Environmental Microbiology*, vol 68, pp1846–1853

Mazzola, M. (2002) 'Mechanisms of natural soil suppressiveness to soilborne diseases', *Antonie van Leeuwenhoek*, vol 81, pp557–564

Mazzola, M. (2004) 'Assessment and management of soil microbial community structure for disease suppression', *Annual Review of Plant Pathology*, vol 42, pp35–59

Milling, A., Smalla, K., Maidl, F. X., Schloter, M. and Munch, J. C. (2004) 'Effects of transgenic potatoes with an altered starch composition on the diversity of soil and rhizosphere bacteria and fungi', *Plant and Soil*, vol 266, pp23–39

Moreira, F. M. S. and Siqueira, J. O. (2002) *Microbiologia e bioquímica do solo,* Editora UFLA, Lavras, Brazil

Muyzer, G. and Smalla, K. (1998) 'Application of denaturing gel electrophoresis (DGGE) and temperature gradient gel electrophoresis (TGGE) in microbial ecology', *Antonie van Leeuwenhoek*, vol 73, pp127–141

Muyzer, G., de Waal, E. C. and Uitterlinden, A. G. (1993) 'Profiling of complex microbial populations by denaturing gradient electrophoresis analysis of polymerase chain reaction-amplified genes coding for 16S rRNA', *Applied and Environmental Microbiology*, vol 59, pp695–700

Nechwatal, J., Schlenzig, A., Jung, T., Cooke, D. E. L., Duncan, J. M. and Osswald, W. F. (2001) 'A combination of baiting and PCR techniques for the detection of *Phytophthora quercina* and *P. citricola* in soil samples from oak stands', *Forest Pathology*, vol 31, pp85–97

Nikolcheva, L. G. and Bärlocher, F. (2004) 'Taxon-specific fungal primers reveal unexpectedly high diversity during leaf decomposition in a stream', *Mycological Progress*, vol 3, pp41–49

O'Donnell, A. G., Goodfellow, M. and Hawksworth, D. L. (1994) 'Theoretical and practical aspects of the quantification of biodiversity among microorganisms', *Philosophical Transactions Royal Society London, B. Biological Sciences*, vol 345, pp65–73

O'Donnell, K. and Cigelnik, E. (1997) 'Two divergent intragenomic rDNA ITS2 types within a monophyletic lineage of the fungus *Fusarium* are nonorthologous', *Molecular Phylogenetics and Evolution*, vol 7, pp103–116

Oros-Sichler, M., Gomes, N. C. M., Neuber, G. and Smalla, K. (2006) 'A new semi-nested PCR protocol to amplify large 18S rRNA gene fragments for PCR-DGGE analysis of soil fungal communities', *Journal of Microbiological Methods*, vol 6, pp63–75

Papavizas, G. C., Adams, P. B., Lomsden, R. D., Lews, J. A., Dow, R. L., Ayers, W. A. and Kantzer, J. G. (1975) 'Ecology and epidemiology of *Rhizoctonia solani* in field soil', *Phytopathology*, vol 65, pp871–877

Persiani, A. M., Maggi, O., Casado, M. A. and Pineda, F. D. (1998) 'Diversity and variability in soil fungi from a disturbed tropical rain forest', *Mycologia*, vol 90, pp206–214

Pettitt, T. R., Wakeham, A. J., Wainwright, M. F. and White, J. G. (2002) 'Comparison of serological, culture, and bait methods for detection of *Pythium* and *Phytophthora* zoospores in water', *Plant Pathology*, vol 51, pp720–727

Pfenning, L. H. (1997) 'Soil and rhizosphere microfungi from Brazilian tropical forest ecosystems', in K. D. Hyde (ed) *Biodiversity of Tropical Microfungi*, Hong Kong University Press, Hong Kong, pp341–365

Ranjard, L., Poly, F., Lata, J. C., Mouguel, C., Thioulouse, J. and Nazaret, S. (2001) 'Characterization of bacterial and fungal soil communities by automated ribosomal intergenic spacer analysis fingerprints, biological and methodological variability', *Applied and Environmental Microbiology*, vol 67, pp4479–4487

Rodrigues, K. F. (1994) 'The foliar fungal endophytes of the Amazonian palm *Euterpe oleraceae*', *Mycologia*, vol 86, pp376–385

Rodrigues-Guzman, M. P. (2001) 'Biodiversidad de los hongos fitopatógenos del suelo de Mexico', *Acta Zoológica Mexicana*, special issue no 1, pp53–78

Rossman, A. Y., Tulloss, R. E., O'Dell, T. E. and Thorn, R. G. (1998) *Protocols for an All Taxa Biodiversity Inventory of Fungi in a Costa Rican Conservation Area*, Parkway Publishers, Boone, NC

Ryan, M. J. and Smith, D. (2004) 'Fungal genetic resource centres and the genomic challenge', *Mycological Research*, vol 108, pp1351–1362

Sanfuentes, E., Alfenas, A. C., Maffia, L. A. and Silveira, S. F. (2002) 'Comparison of baits to quantify inoculum density of *Rhizoctonia* spp. in Eucalyptus clonal garden soils', *Australasian Plant Pathology*, vol 31, pp177–183

Schneider, R. W. (1984) *Suppressive Soils and Plant Disease*, APS Press, St Paul, MN

Schuessler, A., Schwarzott, D. and Walker C. (2001) 'A new fungal phylum, the Glomeromycota: Phylogeny and evolution', *Mycological Research*, vol 105, pp1413–1421

Selenska, S. and Klingmüller, W. (1992) 'Direct recovery and molecular analysis of DNA and RNA from soil', *Microbial Releases*, vol 1, pp41–46

Setälä, H. and McLean, M. A. (2004) 'Decomposition rate of organic substrates in relation to the species diversity of soil saprophytic fungi', *Oecologia*, vol 139, pp98–107

Siebert, P. D. and Larrick, J. W. (1992) 'Competitive PCR', *Nature*, vol 359, pp557–558

Silva, M., Umbuzeiro, G. A., Pfenning, L. H., Canhos, V. P. and Esposito, E. (2003) 'Filamentous fungi isolated from estuarine sediments contaminated with industrial discharges', *Soil and Sediment Contamination*, vol 12, pp345–356

Singh, B. K., Munro, S., Reid, E., Ord, B., Potts, J. M., Paterson, E. and Millard, P. (2006) 'Investigating microbial community structure in soils by physiological, biochemical and molecular fingerprinting methods', *European Journal of Soil Science*, vol 57, pp72–82

Singleton, L. L., Mihail, J. D. and Rush, C. M. (1992) *Methods for Research on Soilborne Phytopathogenic Fungi*, APS Press, St Paul, MN

Smit, E., Leeflang, P., Glandorf, B. and van Elsas, J. D. (1999) 'Analysis of fungal diversity in the wheat rhizosphere by sequencing of cloned PCR-amplified genes encoding 18S rRNA and temperature gradient gel electrophoresis', *Applied and Environmental Microbiology*, vol 66, pp2614–2621

Smith, D. and Waller, J. M. (1992) 'Culture collections of microorganisms: Their importance in tropical plant pathology', *Fitopatologia Brasileira*, vol 17, pp5–12

Sneh B., Burpee, L. and Ogoshi, A. (1991) *Identification of Rhizoctonia Species*, APS Press, St Paul, MN

Sokal, R. R. and Rohlf, J. F. (1995) *Biometry: The Principles and Practice of Statistics in Biological Research*, 3rd edition, W. H. Freeman, New York

Thorn, R. G., Reddy, C. A., Harris, D. and Paul, E. A. (1996) 'Isolation of saprophytic basidiomycetes from soil', *Applied Environmental Microbiology*, vol 62, pp4288–4292

Tsai, Y. L. and Olson, B. H. (1992) 'Rapid method for direct extraction of DNA from soil and sediments', *Applied Environmental Microbiology*, vol 58, pp1070–1074

Tsao, P. H., Erwin, D. C. and Bartnicki-Garcia, S. (1983) *Phytophthora, its Biology, Taxonomy, Ecology and Pathology*, APS Press, St Paul, MN

Vandenkoornhuyse, P., Baldauf, S., Leyval, C., Straczek, J. and Young, J. P. W. (2002) 'Extensive fungal diversity in plant roots', *Science*, vol 295, p2051

Viaud, M., Pasquier, A. and Brygoo, Y. (2000) 'Diversity of soil fungi studied by PCR-RFLP of ITS', *Mycological Research*, vol 104, pp1027–1032

Wainwright, M. (1988) 'Metabolic diversity of fungi in relation to growth and mineral cycling in soil: A review', *Transactions of the British Mycological Society*, vol 90, pp159–170

Wellington, E. M. H., Berry, A. and Krsek, M. (2003) 'Resolving functional diversity in relation to microbial community structure in soil, exploiting genomics and stable isotope probing', *Current Opinion in Microbiology*, vol 6, pp295–301

Whipps, J. M., McQuilken, M. P. and Budge, S. P. (1993) 'Use of fungal antagonists for biocontrol of damping-off and *Sclerotinia* disease', *Pesticide Science*, vol 37, pp309–317

Widden, P. and Parkinson, D. (1973) 'Fungi from Canadian coniferous forest soils', *Canadian Journal of Botany*, vol 51, pp2275–2290

Yergeau, E., Filion, M., Vujanovic, V. and St-Arnaud, M. (2005) 'A PCR-denaturing gradient gel electrophoresis approach to assess *Fusarium* diversity in asparagus', *Journal of Microbiological Methods*, vol 60, pp143–154

Zak, J. C. and Visser, S. (1996) 'An appraisal of soil fungal diversity: The crossroads between taxonomic and functional biodiversity', *Biodiversity and Conservation*, vol 5, pp169–183

Handbooks and manuals for identification

Barnett, H. L. and Hunter, B. E. (1998) *Illustrated Genera of Imperfect Fungi*, 4th edition, APS Press, St Paul, MN

Bissett, J. (1984) 'A revision of the genus *Trichoderma*, I. Section *Longibrachiatum* sect. nov.', *Canadian Journal of Botany*, vol 62, pp924–931

Bissett, J. (1991a) 'A revision of the genus *Trichoderma*, II. Infrageneric classification', *Canadian Journal of Botany*, vol 69, pp2357–2372

Bissett, J. (1991b) 'A revision of the genus *Trichoderma*, III. Section *Pachybasium*', *Canadian Journal of Botany*, vol 69, pp2373–2417

Bissett, J. (1991c), 'A revision of the genus *Trichoderma*, IV. Additional notes on section *Longibrachiatum*', *Canadian Journal of Botany*, vol 69, pp2418–2420

Crous, P. W. (2002b) *Taxonomy and pathology of Cylindrocladium (Calonectria) and Allied Genera*, APS Press, St Paul, MN

Domsch, K. H., Gams, W. and Anderson, T. (2007) *Compendium of Soil Fungi, 2nd edition*, IHW Verlag, Eching

Ellis, M. B. (1971) *Dematiaceous Hyphomycetes*, CMI, Kew, London

Ellis, M. B. (1976) *More Dematiaceous Hyphomycetes*, CMI, Kew, London

Ellis, M. B. and Ellis, J. P. (1985) *Microfungi on Land Plants: An Identification Handbook*, expanded reprint (Richmond Publ. Co. 1997), Croom Helm, London

Erwin, D. C. and Ribeiro, O. L. K. (1996) *Phytophthora Diseases Worldwide*, APS Press, St Paul, MN

Gams, W., Hoekstra, E. S. and Aptroot, A. (1998) *CBS Course of Mycology*, 4th edition, CBS, Baarn, The Netherlands

Hanlin, R.T. 1990. *Illustrated Genera of Ascomycetes*, vol I, American Phytopathological Society, St Paul, MN

Hanlin, R. T. (1998a) *Illustrated Genera of Ascomycetes*, vol II, American Phytopathological Society, St Paul, MN

Hanlin, R. T. (1998b) *Combined Keys to Illustrated Genera of Ascomycetes*, vols I and II, American Phytopathological Society, St Paul, MN

Karling, J. S. (1977) *Chytridiomycetarum Iconographia*, J. Cramer, Vaduz, Liechtenstein

Kirk, P. M., Cannon, P. F., David, J. C. and Stalpers, J. A. (2001) *Dictionary of the Fungi*, 9th edition, CAB International, Wallingford

Klich, M. A. and Pitt, J. I. (1988) *A Laboratory Guide to Common Aspergillus Species and Their Teleomorphs*, CSIRO, North Ryde, Australia

Leslie, J. F., Summerell, B. A. and Bullock, S. (2006) *The Fusarium Laboratory Manual*, Blackwell Publishers, Malden, MA

Nag Raj, T. R. (1993) *Coelomycetous Anamorphs with Appendage-bearing Conidia*, Mycologue Publications, Waterloo, ON

Pitt, J. I. (2000) *A Laboratory Guide to Common Penicillium Species*, 3rd edition, CSIRO, North Ryde, Australia

Plaats-Niterink, A. J. van der (1981) Monograph of the genus *Pythium*, *Studies in Mycology*, vol 21, pp51–53

Rondon, M. R., August, P. R., Betterman, A. D., Brady, S. F., Grossman, T. H., Liles, M. R., Loiacono, K. A., Lynch, B. A., MacNeil, I. A., Minor, C., Tiong, C. L., Gilman, M., Osburne,

M. S., Clardy, J., Handelsman, J., Goodman, R. M. (2000) 'Cloning the soil metagenome: A strategy for accessing the genetic and functional diversity of uncultured microorganisms', *Applied Environmental Microbiology*, vol 66, pp2541–2547
Samson, R. A., Hoekstra, E. S., Frisvad, J. C. and Filtenborg, O. (2000) *Introduction to Food and Air-borne Fungi*, 6th edition, CBS, Baarn, The Netherlands
Sneh, B., Burpee, L. and Ogoshi, A. (1991) *Identification of Rhizoctonia Species*, APS Press, St Paul, MN
Sutton, B. C. (1980) *The Coelomycetes*, CMI, Kew, London
Tsao, P. H., Erwin, D. C. and Bartnicki-Garcia, S. (1983) *Phytophthora: Its Biology, Taxonomy, Ecology and Pathology*, APS Press, St Paul, MN
Waterhouse, G. M. (1967) 'A key to *Pythium* Pringsheim', *Mycological Papers*, vol 109, pp1–15
Waterhouse, G. M. (1968) 'The genus *Pythium* Pringsheim: Diagnoses or descriptions and figures from the original papers', *Mycological Papers*, vol 110, pp1–50
Waterhouse, G. M. (1970) 'The genus *Phytophthora* de Bary: Diagnoses or descriptions and figures from the original papers', *Mycological Papers*, vol 122, pp1–59

Websites

Johnson, T. W., Jr, Seymour, R. L. and Padgett, D. E. (2002) 'Biology and systematics of Saprolegniaceae', www.uncw.edu/people/padgett/book, accessed November 2002
Samuels, G. J., Chaverri, P., Farr, D. F. and McCray, E. B. (undated) 'Trichoderma Online', Systematic Botany & Mycology Laboratory, ARS, USDA, http://nt.ars-grin.gov/taxadescriptions/keys/TrichodermaIndex.cfm

Appendix 8.1 Culture media and additives: Baits

Recipes are for one litre, based on Gams et al (1998) and Erwin and Ribeiro (1996).

MA2% – malt extract agar
malt extract 20g
agar 18g

SNA – synthetic nutrient-poor agar
KH_2PO_4 1.0g
KNO_3 1.0g
$MgSO_4 \times 7H_2O$ 0.5g
KCl 0.5g
glucose 0.2g
saccharose 0.2g
agar 18g

OA – oatmeal agar
oat flakes 30g
agar 15g

PDA – potato dextrose agar
potato, diced 200g
dextrose 20g
agar 18g

PCA – potato carrot agar
potato 20g
carrot 20g
agar 18g

CA – carrot agar
carrot, ground 100g
agar 18g

CA + p.p.s
carrot, ground 100g
agar 18g
penicillin 0.2g
pimaricin 0.02g
streptomycin 0.1g

V8 agar
V8 100mL
$CaCO_3$ 2.0g
agar 18g
distilled water 900mL

PARP
cornmeal 17g
pimaricin 10mg
ampicillin 250mg
rifampicillin 10mg
pentachloronitrobenzene 100mg
hymexazol 50mg

CMA+p.p.s. – cornmeal agar
cornmeal 8.0g
penicillin 0.2g
pimaricin 0.02g
streptomycin 0.1g
agar 18g

MP5 – maltose peptone agar
maltose 4.0g
peptone 1.0g
agar 18g

Antibiotics, if needed
penicillin 0.2g
pimaricin 0.02g
streptomycin 0.1g

Baits

The use of baits is imperative for recovering zoospore-forming fungi from complex substrates like soil. Depending on what fungus is targeted specific baits can be used. Vegetable baits should be washed; those of animal origins are sterilized using UV light for one hour.

Cellulosic substrates: *Sorghum* seeds, cellophane paper, dried maize leaves, epidermis of onion, pollen grains of *Pinus*, pieces of apple, potato, carrot or cucumber.

Chitinous substrates: snake skin, exoskeleton of shrimps, insect wings.

Keratinous substrates: blonde child's hairs.

Chapter 9

Sampling, Conserving and Identifying Fruit Flies

Neliton Marques da Silva

Introduction

Among the fruit pests of tropical America, the fruit flies are considered of greatest economic importance, as they are key pests for most of the fruit crops. They are multivoltine insects with relatively high biotic potential, and great capacity to infest different native and exotic fruit species. They belong to the order Diptera, family Tephritidae. Five genera are important as pests: *Anastrepha*, *Bactrocera*, *Ceratitis*, *Rhagoletis* and *Dacus*, which are spread globally throughout the continents, except Antarctica (White and Elson-Harris, 1992). A few studies have highlighted the ecology and aetiological aspects of fruit flies, mainly focusing on the pupal and larval phases (Silva et al, 1996; Zucchi et al, 1996). Fruit flies have complex behaviour and taxonomy (Bateman, 1972; Steck and Wharton, 1988). Classification is exclusively based on adult morphological characteristics. The sexes are easily distinguished, as females have an ovipositor quite prominent at the end of the abdomen, with a long and fine tip. Taxonomic characteristics differentiating gender in larvae and pupae are not yet established (Salles, 2000).

Damage is caused during the immature phase, a period in which the larvae destroy the fruit pulp, making them unsuitable for harvesting and consumption. Before they reach the adult stage, the larvae migrate from the fruit to pupate in the soil (Plate 8a). Thus, fruit flies are temporary soil inhabitants, because they live only part of their life cycle in soil. After infested fruits have dropped, larvae move on the soil surface to find suitable soil conditions, penetrate to approximately 10cm depth and pupate. This depth can vary according to physical soil conditions, mainly temperature, humidity and texture. The entire pupal stage occurs in soil. Therefore, for a part of their life cycle, fruit flies can be considered soil organisms. The pupal phase lasts from eight to ten days depending on temperature and humidity. The pupal phase is the one most vulnerable to abiotic stress and to many natural enemies, mainly predators and entomopathogens such as bacteria, fungi and nematodes which are also part of the soil biota (see Chapter 10). These antagonists play an important role in the biological control of fruit flies. Even if larvae are parasitized (infected) inside the fruits, the adult parasites can emerge in the soil. Therefore, it is of fundamental importance to consider soil biodiversity as a factor influencing larval and pupal mortality when designing integrated pest management strategies.

Because the life cycle involves the aerial parts of the plant host as well the soil, sampling procedures vary according to the stage and purpose. For the adult fly, which is diurnal and responds to visual and olfactory stimuli, the use of traps containing food baits is most suitable. However, this technique does not unambiguously establish the relationship between the fly and its host. This can only be achieved by collecting infested fruits, branches or seeds, where the fruit fly larvae are lodged.

Sampling fruit flies

Sampling of adult fruit flies is carried out by using plastic MacPhail traps (Plate 8b) containing a food bait, usually 200ml of hydrolyzed corn protein (5 per cent in water preserved with sodium tetraborate, at pH between 8.5 and 9.0). Alternatively, it is possible to use fruit juice at 10 per cent, sugar at 10 per cent or sugar cane syrup at 10 per cent. The food baits should be replaced weekly and trapped specimens removed. The traps are installed in the middle of the tree canopy and the location is georeferenced (GPS). The number of traps per unit area can vary according to project objectives. If the objective is pest monitoring, one trap per 4ha is sufficient. When the objective is integrated control, the density should be three to five traps per ha or according to the extant land use systems, that is, all land uses represented and monitored with the same trap density. Up to five traps per ha can be installed in areas with shrubs or trees (home gardens), while up to three per ha are suitable in herbaceous (arable) plantings. In the case of herbaceous plants the trap can be supported by three sticks positioned to form a tripod-like structure, or placed in an adjacent tree if one is available. Traps must be positioned equidistant from each other.

Collecting the trapped adults

As several insect species of diverse taxonomic categories are likely to be trapped, it is first necessary to sort out fruit flies from other specimens. This can be done in the field or in the lab. When food baits are replaced, the contents of the trap should be passed through a fine nylon mesh of 1.5mm to remove the flies. Using curved forceps, the fruit flies are sorted out from other taxonomic groups. Their sex is recorded and they are stored in labelled glass jars (around 50ml), containing 70 per cent alcohol for further identification. The label must include basic sampling information and the trap number. For example:

- Manaus-AM
- Brazil
- 04° 05'S; 60° 04'W
- 23 March 2006
- Silva, N. M.
- Trap No 5

Other taxonomic groups collected in the traps should also be noted.

Fruit sampling

To establish the relationships of the *Anastrepha* species with their/its host(s), fruits are collected randomly in different land use systems and at different stages of maturation. The fruits must be collected directly from the trees or immediately after they fall to the soil. Fruits sampled are separated by species, protected in cloth bags and transported in insulated cool boxes to the lab. After that, they are weighed, counted and separated by sampling point into plastic trays containing a layer of vermiculite or fine sand, which serves as substratum for pupation. Finally, the trays are then covered by a voile cloth, secured tightly with rubber bands to prevent the escape of any adult flies that emerge.

Obtaining the pupae

The substratum of vermiculite or fine sand is passed through a galvanized iron mesh of 1.5mm to separate the pupae, which are placed in cages with the respective identification labels, to allow the emergence of the adult fruit flies and/or parasitoids. The cages should be examined daily. Emerged adults are retained in the cages for 48 hours, to allow cuticular hardening and the full development of the wing spots (Plate 8c), which are of great importance for taxonomic identification.

After emergence adult flies are fed with a 10 per cent aqueous solution of honey, changed daily. Dates of emergence and the number and sex of adult fruit flies or parasitoids must be recorded. Finally, specimens are fixed in 70 per cent alcohol. The determination of sexual ratio (SR) (i.e. females to males ratio), for both adult fruit flies and parasitoids, is made according to Silveira Neto et al (1976) using the equation below:

$$SR = \frac{\text{No of females}}{\text{No of females} + \text{No of males}}$$

Species identification

The taxonomic identification of tephritid species is based on the ventral examination of the female apical aculeus, under a stereoscopic microscope (40×) or in mounted slides for examination in the transmission microscope (100×). It is first necessary to secure the extroversion of the aculeus with the help of needles or forceps (Plate 8d), as described by Zucchi (1988). Species identification within *Anastrepha* is based on the adult females, by reference to the wing pattern, body colour, mesonotum, mediotergite, abdomen and especially the morphological characteristics of the apical aculeus (Plate 8e), which are compared with museum specimens and submitted to taxonomic keys according to Lima (1934), Stone (1942), Foote (1967), Steyskal (1977), Silva (1993), Ronchi-Teles (2002), Zucchi (1978, 2000) and Norrbom (1985). It is recommended that voucher species are deposited at a museum or added to the collection of the institutional laboratory.

REFERENCES

Bateman, M. A. (1972) 'The ecology of fruit flies', *Annual Review of Entomology*, vol 17, pp493–518

Foote, R. H. (1967) 'Family Tephritidae', in M. Vanzolini (ed) *A Catalogue of the Diptera of the Americas South of the United States*, Secretaria da Agricultura, Departamento de Zoologia, São Paulo

Lima, A. C. (1934) 'Moscas-de-frutas do gênero *Anastrepha* Schiner, 1868 (Diptera: Trypetidae)', *Memórias do Instituto Oswaldo Cruz (Rio de Janeiro)*, vol 28, issue 4, pp487–575

Norrbom, A. L. (1985) 'Phylogenetic analysis and taxonomy of the *cryptostrepha*, *daciformis*, *robusta* and *schausi* species groups of *Anastrepha* Schiner (Diptera: Tephritidae)', PhD Thesis, The Pennsylvania State University, Pennsylvania

Ronchi-Teles, B. (2002) 'Ocorrência e flutuação populacional de espécies de moscas-das-frutas e parasitóides com ênfase para o Gênero *Anastrepha* (Diptera: Tephritidae) na Amazônia Brasileira', PhD Thesis, Federal University of the Amazon (UFAM) and National Institute for Agricultural Research (INPA), Manaus, Brazil

Salles, L. A. (2000) 'Biologia e ciclo de vida de *Anastrepha fraterculus* (Wied.)', in A. Malavasi and R. A. Zucchi (eds) *Moscas-das-frutas de importância econômica no Brasil: Conhecimento básico e aplicado*, Ribeirão Preto, Holos, Brazil

Silva, N. M. (1993) 'Levantamento e análise faunística de moscas-das-frutas (Diptera: Tephritidae) em quatro locais do Estado do Amazonas', PhD Thesis, Escola Superior de Agricultura 'Luiz de Queiroz', University of São Paulo, São Paulo

Silva, N. M., Zucchi, R. A., Silveira Neto, S. (1996) 'The natural hosts plants of *Anastrepha* (Diptera: Tephritidae) in the State of Amazonas, Brazil', in G. J. Steck and B. A. MacPherson (eds) *Fruit Fly Pests: A World Assessment of their Biology and Management*, CRC Press, Boca Raton

Silveira Neto, S., Nakano, O., Barbin, D. and Nova, N. A. V. (eds) (1976) *Manual de Ecologia dos Insetos*, Ceres, São Paulo

Steck, J. and Wharton, R. A. (1988) 'Description of immature stages of *Anastrepha interrupta*, *A. limae*, and *A. grandis* (Diptera: Tephritidae)', *Annals of the Entomological Society of America*, vol 81, issue 6, pp994–1003

Steyskal, G. C. (1977) *Pictorial Key to Species of the Genus* Anastrepha *(Diptera: Tephritidae)*, Entomological Society of Washington, Washington, DC

Stone, A. (1942) 'The fruit flies of the genus *Anastrepha*', *United States Department of Agriculture Miscellaneous Publications*, vol 439, pp1–112

White, I. M. and Elson-Harris, M. M. (1992) *Fruit Flies of Economic Significance: Their Identifications and Bionomics*, CAB International, Wallingford

Zucchi, R. A. (1978) 'Taxonomia das espécies de *Anastrepha* Schiner 1868 (Diptera: Tephritidae) assinaladas no Brasil', PhD Thesis, Escola Superior de Agricultura 'Luiz de Queiroz', University of São Paulo, São Paulo

Zucchi, R. A. (1988) 'Moscas-das-frutas (Dip., Tephritidae) no Brasil: taxonomia, distribuição geográfica e hospedeiros', in H. L. M. de Souza (ed) *Moscas-das-frutas no Brasil*, Fundação Cargill, Campinas, Brazil

Zucchi, R. A. (2000) 'Taxonomia', in A. Malavasi and R. A. Zucchi (eds) *Moscas-das-frutas de importância econômica no Brasil: Conhecimento básico e aplicado*, Holos, Ribeirão Preto, Brazil

Zucchi, R. A., Silva, N. M. and Silveira Neto, S. (1996) '*Anastrepha* species (Diptera: Tephritidae) from the Brazilian Amazon: Distribution, hosts and lectotype designations', in G. J. Steck and B. A. MacPherson (eds) *Fruit Fly Pests: A World Assessment of their Biology and Management*, CRC Press, Boca Raton

Chapter 10

Entomopathogenic Fungi and Nematodes

Alcides Moino, Jr and Ricardo Sousa Cavalcanti

INTRODUCTION

There are about half a million described species of insects on the Earth, and the true diversity may be many times this figure (Groombridge, 1992). Approximately 10 per cent of these can be considered agricultural, forestry or urban pests. If we assume that each insect species is susceptible to at least one pathogenic microorganism, often host-specific, we have an insight into the potential importance of the study of these pathogens in the context of pest control and biodiversity.

Insect pathology is the science that studies insect diseases, aiming to use them for control of pest species or with the objective of preventing their occurrence in useful insects. Disease is a dynamic process, in which the host (insect) and the pathogen (microorganism) are both adapted morphologically and physiologically, the former for the infection process and the latter for resistance. Microbial control is a form of biological control that deals with the rational use of the entomopathogens, aiming not to eliminate the pest populations but to maintain them at levels below the economic damage threshold. This mimics the natural dynamics between pathogen and host in the field without human interventions. The principal microorganisms used or of potential use in the microbial control of insects are fungi, bacteria, viruses, nematodes and protozoans.

Many hundreds of species of entomopathogenic fungi are known, attacking a wide range of insects and mites, with varying degrees of host-specificity (Hajek and St Leger, 1994; Roy et al, 2006). The fungi produce spores which germinate on contact with the host and invade the body, killing the host between four and ten days later. After death, many thousands of new spores are produced which disperse and continue the fungal life cycle on new hosts. A small number of generalist species, for example *Lecanicillium logisporum*, which can be readily cultured and therefore mass produced, have been developed as biopesticides for inundative use on pest populations, but this technology has proved disappointing in practice because of high cost, poor persistence (especially under tropical conditions) and low efficacy in comparison with chemical (i.e. toxin-based) insecticides.

Entomopathogenic nematodes are about 0.5mm in length. Juvenile nematodes parasitize their hosts by directly penetrating the cuticle or through natural openings such as spiracles; bacteria introduced with the parasite increase rapidly and kill the host, allowing the nematodes to grow and mature on the decomposing tissue, and reproduce as adults. A new generation of infective juveniles emerges between one and two weeks after the initial invasion of the host (Kaya and Gaugler, 1993). Two family-level groups, the steinernematids and the heterorhabditids, are obligate parasites of insects and have been made the basis of a number of biological pesticides designed particularly for use against soil pests such as weevils and fly larvae, but again there are disadvantages related to high cost, low persistence and (for nematodes) inactivity at low temperatures.

Despite the limited success of entomopathogenic fungi and nematodes as biocidal products, they appear to be both diverse and universal components of soil biotas, where they may be exceptionally virulent (i.e. causing rapid death of the host and high mortality in host populations) and cause periodic epizootics (Chandler et al, 1997; Myers and Rothaman, 1995). Most of the pathogens have a 'sit and wait' transmission strategy (sensu Ewald, 1995): the organisms produce infective stages which are released into the environment when the host dies and have the ability to enter diapause or remain dormant until new hosts become available. This is adaptive, as their much larger arthropod hosts are typically patchy in distribution, with a geographically widespread metapopulation and high mobility. Diapause and dormancy decreases dependence on host mobility and may be supplemented by the ability to live saprotrophically in soil, though competitive efficiency against other free-living saprotrophs may be low. Although above-ground arthropods are susceptible to fungi and nematodes as well as below-ground ones, the soil is clearly a reservoir of the infective stages, because it offers a stable micro-environment with a suitable pore structure for nematodes and organic resources for fungi. There is often no correlation between host density and disease occurrence, again suggesting that the pathogens are very widespread in the soil; in the case of entomopathogenic fungi this is supported by molecular evidence that soil populations are not wholly clonal, despite the absence of overt sexual stages, and that significant gene flow occurs locally (e.g. Bidochka et al, 2001).

Biology of Entomopathogenic Fungi and Nematodes

Fungi

Fungi can be unicellular microorganisms (yeasts), or multicellular (filamentous) consisting of elongated cells provided with a wall containing cellulose and chitin together with other carbohydrates and proteins. These vegetative structures are called hyphae. After the successful infection of a host individual, reproductive structures known as spores or conidia are produced asexually, through which the pathogen is disseminated. Fungi possess great genetic variability and a wide range of hosts can be attacked by the group as a whole. The main fungi of interest as insect pathogens are: *Beauveria bassiana*, *Metarhizium anisopliae*, *M. flavoviride*, *Nomuraea rileyi*, *Lecanicillium lecanii*, *Hirsutella thompsonii*, *Aschersonia aleyrodis*, *Paecilomyces* spp., *Cordyceps* spp., and fungi from the order Entomophthorales (*Zoophthora*, *Entomophthora*, *Entomophaga*, *Neozygites*).

Nematodes

Entomopathogenic nematodes are acoelomate roundworms, very similar to those that infect plants, and that can be associated with insects in three ways: phoresy (passive adherence and transport), obligate parasitism and facultative parasitism. In general, entomopathogenic nematodes have a very close association (symbiosis) with particular bacteria which are the primary agents in the establishment of the infection in the host. The nematodes carry specific bacteria internally, and these are set free in the interior of the insect body after the nematode penetrates through natural openings such as the mouth, spiracles or anus. These bacteria then multiply in the insect, killing it by septicaemia (generalized infection). The main nematode groups of interest belong to the order Rhabditida, and the families Steinernematidae and Heterorhabditidae.

The correct identification of an entomopathogenic agent is based (for fungi) on morphological characters of the organism in culture and the structure of conidiophores and conidia and (for nematodes) on measurements made on infective juvenile structures. However, to do this it is necessary to isolate the microorganism in pure cultures or at least to isolate populations. The microbial population in any natural environment is very large and therefore, when a dead insect is collected for the culture of infective inoculum, it is very common also to find fungi, bacteria and other saprotrophic microorganisms that have no potential as specific pest-control agents. Similarly, when the aim of the sampling is to evaluate biodiversity, as for example in soil, there will be a similar result, and additional measures will be required to identify prospective pathogens. We describe two basic techniques for isolation of entomopathogenic fungi and nematodes from soil samples.

Methodology for isolation of entomopathogenic fungi (Chase et al, 1986; Liu et al, 1993; Alves et al, 1998a)

Soil samples are collected at 0–20cm depth, and placed in labelled plastic bags. At each site, a composite sample is made from six subsamples collected at points located around the central monolith (see Chapter 2). Samples can be conveniently taken by coring (see Chapter 4, Mesofauna), although this is not specified as essential. In the laboratory the composite sample is mixed thoroughly and final aliquots of 1g of soil are taken. Tenfold dilutions of the aliquots are prepared successively using sterilized distilled water until a 10,000-fold dilution is reached. From each sample replicates of 0.1mL of the 10^3 and 10^4 dilutions are used to inoculate a selective culture medium (containing the fungicide dodine) and also PDA medium (potato dextrose agar). The selective medium with dodine is made as follows: 20.0g oat flour + 1 litre distilled water; sterilize for 20 minutes at 120°C and filter; add distilled water to bring the volume back to one litre and add 20.0g agar, 550mg dodine (N-dodecilguanidine acetate), 5.0mg tetracycline, 10.0mg crystal violet. Potato dextrose agar is prepared as follows: 200g potato infusion (potatoes diced and boiled to extract starch), 20g dextrose, 15g agar and distilled water to bring the total volume to one litre.

The fungicide dodine is added in small concentrations (close to 10mg mL^{-1}) to the composition of the selective medium to isolate entomopathogenic fungi from the soil

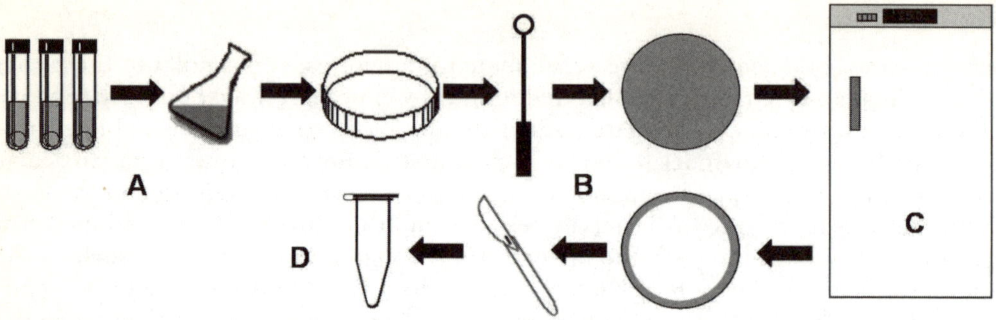

Note: Separation from contaminants was achieved by collecting a small portion of any of the fungal colonies of interest with a needle and transferring this portion to a fresh Petri dish with PDA culture medium (three inoculation points per dish).

Figure 10.1 *Procedures to isolate entomopathogenic fungi on culture media: a) serial dilution of the fungal aqueous suspension; b) inoculation of the suspension (0.1mL aliquot) onto Petri dish with culture medium; c) incubation under controlled conditions in BOD chamber; d) permanent freezer storage of conidia in Eppendorf tubes*

and to preserve the isolates least susceptible to the fungicide. This is effective especially in relation to *Metarhizium anisopliae*. After this, 0.1mL from the last dilution is inoculated onto the surface of the culture medium in a Petri dish and spread with a Drigalski handle (glass or stainless steel spreader). The dishes are incubated at 27°C (Figure 10.1). Fungal growth and sporulation are evaluated between 7 and 15 days after the inoculation. After incubation, the identification of the cultures of interest (mainly *Metarhizium anisopliae*, *Beauveria bassiana* and *Paecilomyces* spp.) can be done using an optical microscope, and identification keys based on the morphological aspects of the reproductive structures, such as conidiophores, conidia and phialides (Alves et al, 1998b; Samson et al, 1998), with subsequent culture purification (Figure 10.2) and conidial storage in Eppendorf tubes in freezer conditions.

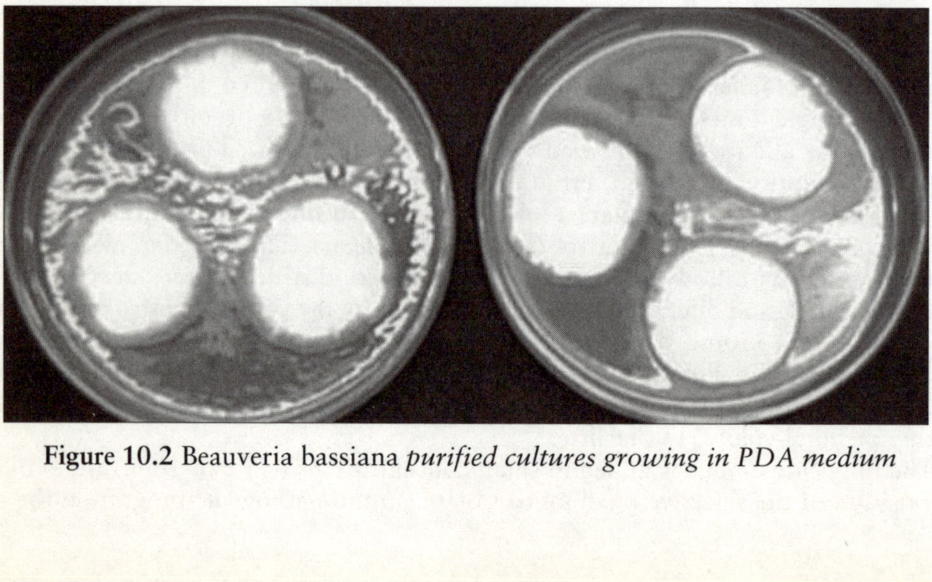

Figure 10.2 Beauveria bassiana *purified cultures growing in PDA medium*

METHODOLOGY FOR ISOLATION OF ENTOMOPATHOGENIC NEMATODES (BEDDING AND ARKHURST, 1974)

Samples are collected in the same way as described above. Detection of nematodes is by bioassay using the insect trap technique with five larvae of *Galleria mellonela* (L.) (Lepidoptera: Pyralidae). The larvae of this lepidopteran are known as mealworms, and can be purchased commercially. The larvae are placed in 500mL plastic boxes together with soil samples and kept closed at 23°C in the dark. Larval mortality is assessed between five and seven days. Dead larvae are placed in a White trap (Chen et al, 2004; Figure 10.3) to collect infective juveniles from the dead bodies. In its original form, a White trap can be constructed from two Petri dishes: a 5cm dish is inverted into a 9cm dish containing sterile water or sterile 'S' medium and covered with a 9cm circle of filter paper dipping down into the fluid at the edges of the lower dish. Infected insect larvae are placed at the centre of the filter paper.

'S' medium consists of a litre of stock solution (0.1M NaCl, 0.05M KH_2PO_4, pH 6.0, 1μM cholesterol), to which is added 26mL of a freshly prepared nutrient solution containing K citrate 0.4M, $CaCl_2$ 0.4M, $MgSO_4$ 0.3M, Na EDTA 0.6mM, $FeSO_4.7H_2O$ 0.3mM, $ZnSO_4.7H_2O$ 0.1mM, $CuSO_4$ 0 1μM.

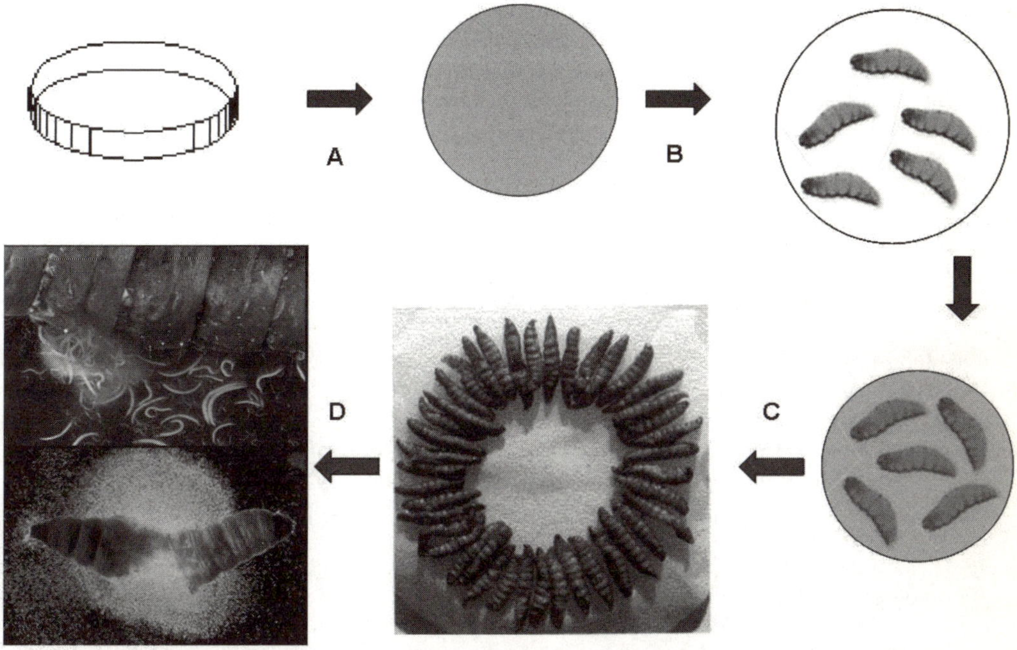

Figure 10.3 *Procedure for the use of the White trap for the isolation of entomopathogenic nematodes: a) Petri dish with filter paper (dry chamber); b) dead* Galleria mellonella *larvae after contact with the soil sample; c) larvae showing typical pathology of infection by nematodes; d) emergence of infective juveniles in the White trap*

After two days, if there are infective juveniles in the fluid, they are collected with distilled water and transferred to a Becker flask. The suspension is kept unstirred until the nematodes settle. The suspension is washed, adding an aqueous solution of formaldehyde (1 per cent) or Ringer's solution, to obtain a final concentration of 10,000 infective juveniles mL^{-1}. Nematodes are stored in closed 50mL containers at or close to 11°C. Identification is made on the basis of specific keys of identification for the appropriate nematode families (Steinernematidae and Heterorhabditidae) (Alves et al, 1998b; Adams and Nguyen, 2002). Specimens from possible new species can be submitted to molecular analyses for comparison of the DNA patterns.

REFERENCES

Adams, B. J. and Nguyen, K. B. (2002) 'Taxonomy and systematics', in R. Gaugler (ed) *Entomopathogenic Nematology*, CAB International, Wallingford

Alves, S. B., Almeida, J. E. M., Moino, A. Jr, and Alves, L. F. A. (1998a) 'Técnicas de laboratório', in S. B. Alves (ed) *Controle Microbiano de Insetos*, 2nd edition, FEALQ, Piracicaba, Brazil

Alves, S. B., Ferraz, L. C. C. B. and Castello Branco A., Jr (1998b) 'Chaves para identificação de patógenos de insetos', in S. B. Alves (ed) *Controle Microbiano de Insetos*, 2nd edition, FEALQ, Piracicaba, Brazil

Bedding, R. A. and Akhurst, R. J. (1974) 'A simple technique for the detection of insect parasitic nematodes in soil', *Nematologica*, vol 21, pp109–110

Bidochka, M. J., Kamp, A. M., Lavender, T. M., Deckoning, J. and De Croos, J. N. A. (2001) 'Habitat association in two genetic groups of the insect-pathogenic fungus *Metarhizium anisopae*: Uncovering cryptic species', *Applied and Environmental Microbiology*, vol 67, pp1335–1342

Chase, A. R. L., Osborne, S. and Ferguson, V. M. (1986) 'Selective isolation of the entomopathogenic fungi *Beauveria bassiana* and *Metarhizium anisopliae* from an artificial potting medium', *Florida Entomologist*, vol 69, pp285–292

Chandler, D., Hay, D. and Read, A. P. (1997) 'Sampling and occurrence of entomopathogenic fungi and nematodes in UK soils', *Applied Soil Ecology*, vol 5, pp133–141

Chen, Z. X., Chen, S. Y. and Dickson, D. W. (2004) *Nematology, Advances and Perspectives*, CAB International, Wallingford

Ewald, P. W. (1995) 'The evolution of virulence: A unifying link between parasitology and ecology', *Journal of Parasitology*, vol 81, pp659–669

Groombridge, B. (ed) (1992) *Global Biodiversity: Status of the Earth's Living Resources*, World Conservation Monitoring Centre, Chapman and Hall, London

Hajek, A. E. and St Leger, R. J. (1994) 'Interactions between fungal pathogens and insect hosts', *Annual Review of Entomology*, vol 39, pp293–322

Kaya, H. K. and Gaugler, R. (1993) 'Entomopathogenic nematodes', *Annual Review of Entomology*, vol 38, pp181–206

Liu, Z. Y., Milner, R. J., McRae, C. F. and Lutton, G. G. (1993) 'The use of Dodine in selective media for the isolation of *Metarhizium* spp. from soil', *Journal of Invertebrate Pathology*, vol 62, pp248–251

Meyers, J. H. and Rothaman, L. E. (1995) 'Virulence and transmission of infectious diseases in humans and insects: Evolutionary and demographic patterns', *Trends in Ecology and Evolution*, vol 10, pp194–198

Roy, H. E., Steinkraus, D. C., Eilenberg, J., Hajek, A. E. and Pell, J. K. (2006) 'Bizarre interactions and endgames: Entomopathogenic fungi and their arthropod hosts', *Annual Review of Entomology*, vol 51, pp331–357

Samson, R. A., Evans, H. C. and Latgé, J. P. (1988) *Atlas for Entomopathogenic Fungi*, Springer-Verlag, New York

Chapter 11

Description and Classification of Land Use at Sampling Locations for the Inventory of Below-ground Biodiversity

E. Jeroen Huising

SUMMARY OF THE METHOD FOR LAND USE DESCRIPTION

This chapter provides a structured list of land use attributes to guide observation of land use characteristics in the field at sampling point locations. The system allows for different levels of detail at which the characteristics are described, depending on the data available or obtainable. In most cases the domains of the attribute values are specified, which requires only the proper value to be assessed in the field. This is done either through direct observation or by interview or inquiries. Only in some cases, generally at the higher level of detail, are actual measurements required. The land use classification is facilitated by a hierarchical structure in which the attributes are ordered, but the final land use classes will depend on the selection of attributes to be considered for the classification, depending on the purpose and context of the classification. How land use classes can be defined with respect to land use intensity is illustrated.

BACKGROUND AND DESIGN PRINCIPLES

Purpose of the land use and land-cover classification

Land use, and in particular the intensity of use, is considered to be one of the determining factors of the abundance and richness of populations of soil organisms, and it is therefore the central hypothesis of the CSM-BGBD project (Giller et al, 2005). To test this hypothesis a system is needed that allows for the consistent recording of land use characteristics at the sampling plot locations and the subsequent classification of land use, across the benchmark areas in the various tropical countries concerned. This chapter presents a structure that provides for the systematic recording of land use characteristics while at the same time providing a structure for the classification of land use. The structure needs to be flexible in the sense that it allows for the use of various inputs, like aerial photos or satellite imagery, for classification purposes and in the sense that it allows for the definition of land use classes based on common denominators across the study sites. The latter implies that the land use classes are not *a priori* defined but depend on the classifiers or the attributes selected for assigning a land use class label to any particular observation, in order to better reflect differences in land use intensity.

The most dominant processes that determine the occurrence and incidence of soil biota and the spatial scale level at which these are manifested are not well known. It is for this reason that a regular grid system was adopted for sampling of below-ground biodiversity, with a detailed inventory of land use and land cover at the sampling locations to allow for analyses of the actual determining factors and *a posteriori* classification. Predefined land use classes may not be applicable (or relevant) to all areas concerned as it will be rather difficult to define a set of standard land use classes given the variation in crops and crop combinations, land use history and land and crop management practices, especially across tropical regions. Also the general notion is that the content of a survey is dependent on the nature of the region (Vink, 1975). Rather than using predefined land use classes the approach described departs from the idea of land use classes being defined based on identification of relevant land use characteristics (attributes), which allows for discrimination between land use classes. The land use classes thus obtained may be used for extrapolation of the findings beyond the study sites.

The detailed description of land use and land cover, therefore, needs to be part of the inventory process. The land use and land-cover classification system provides for the harmonization of procedures for data collection and handling of the data. The aim of the inventory, to provide data for evaluating change in below-ground biodiversity in relation to land use, especially land use intensity, is different from most land use surveys that often have the description of land use and study of the natural and socio-economic conditions of the land use as their prime objective (Vink, 1975). This generally translates into the registration of the occurrence of crops. However, for the present purpose, features related to how the land is being used (i.e. the management of crop and land) are of particular importance since these will greatly impact the distribution of soil organisms. A second objective for the land use inventory is to provide background information for the definition of alternative land uses and management practices that improve sustainability of the agricultural production and conserve below-ground biodiversity.

The above were also the objectives of the Africover programme, which resulted in the development of the Land Cover Classification System (LCCS) by Di Gregorio and Jansen (2000). The approach to land use inventory described here leans heavily on this classification system, though with elements added related to land and crop management. Also, the methods described in this chapter apply only to terrestrial agricultural landscapes (i.e. 'cultivated and managed areas'), whereas the scope of the LCCS includes (semi) natural vegetated terrestrial systems, aquatic systems and primarily non-vegetated areas. For classification and description of (semi) natural vegetation areas the LCCS can be followed. The structure targets description of present land use; no provisions have been made for the recording of land use history. Nor does the method include a structure for the description of environmental attributes, like climate, land form, and so on, though attributes that relate to land use history or environmental conditions could be added as modifiers to the classification system.

The method presented here is intended to be used for the description of land use at plot level. It does not include elements related to livestock management because this is not particularly related to plot level, nor does it include elements related to the farming system, though it is acknowledged that for studies at a larger scale or for considering alternative land use options these need to be considered. The method described here is therefore not suited for the description and mapping of land use of larger areas, other then for description of land cover per se.

Concept and design principles

The concepts and design principles of the LCCS are adopted, meaning that diagnostic criteria are used that are ordered in a hierarchical manner to allow for a consistent classification system, with clear boundary definitions for the classes. Many classification systems use *a priori* defined descriptive land use and land-cover classes, arranged in a hierarchical system with the hierarchies relating to different levels of thematic and spatial detail. See, for example, the US Geological Survey land use and land-cover classification system, as described by Lillesand and Kiefer (1987) based on the interpretation of aerial photographs for land use and land-cover mapping purposes. These classification 'systems' basically represent legends. Such systems often lack clear definition of class boundaries, are rigid with limited capability to incorporate new classes and therefore often non-exhaustive (i.e. they represent a subset of the range of possible land use and cover classes) and because there is no reference to a system of classification, the ability for comparison with other systems is limited.

The basic principle rests on the use of classifiers that refer to independent diagnostic criteria or attributes. Class hierarchy is built on the sets of classifiers used to define a class. The more classifiers added, the more detailed the level of classification. There are five levels in the classification hierarchy:

Level 1 Characteristics related to the main crop and field.
Level 2 Characteristics related to the combination of crops.
Level 3 Characteristics related to the cultural aspects like irrigation and time of cultivation.
Level 4 Characteristics related to land management practices.
Level 5 Land use characteristics of the area directly surrounding the sampling plot, especially in relation to woody vegetation elements not recognized as a crop and in relation to crops grown on adjacent fields and including spatial arrangements of the agricultural fields.

For example, the major class is defined by the type of the main crop (e.g. 'tree crop' for tree based system), together with the size of the fields. The field size generally is indicative of the size of the farm, which is often an indicator of social class and the type of management (Huising, 1993). Sub-classes can be defined based on other crops that form part of the cropping system. That may concern crops grown simultaneously or in sequence. Crop rotation characteristics are considered elsewhere. At the subsequent level of the classification hierarchy cultural practices are considered, relating to water supply characteristics and cultivation timing factors. The latter determines whether the field is part of a shifting cultivation system, a fallow system or is permanently cultivated.

These first three levels provide the basis for the classification of land use, and data collected in relation to these characteristics is suggested to represent the minimum data requirement. We have added classifiers that relate to land and crop management that will allow determination of the level of intensity of use. These classifiers typically refer to operations performed at the particular plot, like land preparation, weeding and fertilizing. The characteristics are not always readily observed, but with some background information on the land use and cropping systems that prevail in the area under consideration, or based on information provided by the farmer, they may be easily assessed.

The fifth level in the classification system relates to the (semi) permanent woody vegetation elements within the plot and surrounding fields; characteristics that are not captured at the previous levels of classification, but that are nevertheless considered important in relation to the intensity of use and that may have a direct impact on below-ground biodiversity. These vegetation elements relate to more permanent features in the landscape, like hedgerows, living fences, wind barriers and other semi-permanent structures.

The other aspect concerns the distribution and arrangement of agricultural fields (i.e. field pattern) and associated crops. This information may assist in determining whether the plot under consideration represents a characteristic (or common) land use and land-cover type and may help to identify possible associations of land uses that may assist in mapping land use (by interpolating from point observations to derive a map that provides a complete cover). Furthermore, the field pattern provides information on land fragmentation, being another aspect of land use intensity with possible direct impact on below-ground biodiversity.

The classification levels define one axis of the classification hierarchy in which additional (sets of) attributes at each subsequent level define sub-classes (representing a higher level of detail). These new attributes provide more specific information on the object under observation. Apart from this axis there is a second axis in the classification hierarchy that refers to the level of precision with which the land use characteristics is described, or that adds specific information (technical details) to the land use characteristic already described. These attributes are therefore not independent and may often involve some kind of measurement at the higher level of detail. For example, the main crop can be described as maize (*Zea mays*), more specific information would refer to the specific variety and this would represent a higher level of detail (specificity). On the other hand the main crop could be described at a more general (genus) level as *Zea* or at family level as Gramineae (grasses). If the class value at the more specific level is known, the class value at the higher level of generalization can be directly inferred. These specific technical attributes, as the LCCS refers to them, can be added without any further consequence for the classification system. The specific technical attributes allow for a more precise definition of associated land use classes that may involve, for example, the quantitative description of land use intensity. The principles of classification hierarchies are explained by Molenaar (1991, 1998) and Huising (1993).

Data recording in the field and additional observations

In the next sections the attribute list for the description of land use is presented. These lists guide the observations in the field at the location of the sampling point and its immediate surroundings. These are therefore plot-level observations. The transformation of the attribute lists into sheets of forms for data recording in the field is straightforward. The possible data values (value domains) for most of the attributes are given to provide for a standard description of land use characteristics. For some of these attributes the value domains are not specified to allow for the definition of data-value classes based on the observed range and distribution of data values. Data classes must provide for a relevant (i.e. meaningful) discrimination of objects and the set of classes should be exhaustive (i.e. non-exclusive). *A priori* defined standard field size classes with fixed class boundaries will not be meaningful for that reason. Attributes can be added to allow for the description of features specific to a particular area, but these should be added at lower levels of the classification hierarchy so that they will not interfere with the classification structure.

Observations for the first three levels of the classification system can be made directly in the field, with information related to the third level (i.e. cultural characteristics) being obtained from an informant or derived from general information on cultural practices in the area. Level 4 type information, related to management, is not obtained from direct observations in the field and will require access to additional sources of information (through interviewing farmers, extension agents or consulting resource documents). However, with knowledge of the prevailing cropping system and management practices in the area one should be able to infer information on the management of the particular plot from direct observation in the field.

Information from these other data sources may not be very reliable or accurate, but this is catered for by applying the hierarchy in the precision of the data (referred to above as the second axis in the classification system). For example, if the information on application rate of a particular inorganic fertilizer is not considered reliable or accurate, it is probably acceptable to classify the application rate as low, medium or high based on the class definition adopted (see Table 11.6b). If that is not the case, the information can be generalized to the class of 'fertilizer applied' (implying the use of inorganic fertilizer). The other way round, with knowledge on common practice in the area and based on secondary observations, a particular application rate (low, medium or high) may be assumed.

The system for the description of land use provides various levels of details. Often the data relating to the most detailed levels of observation are very difficult to obtain or the sources are indeed not reliable. However, this method is not prescriptive in the sense that the data need to be collected at all levels of detail. The level of details at which information is gathered depends on the purpose of the land use inventory and the possibility of obtaining the required information. The system for the description of land use provides a structure that can be implemented to the level of detail considered feasible under the prevailing conditions.

Secondary observations may relate to crop status or presence of weeds, for example. These observations do not qualify as classifiers and are therefore not included in the attribute lists below. However, they could be easily incorporated in the data recording sheets if needed, depending on the experience of the surveyor. See Stocking and Murnaghan (2001) for relevant indicators for observations in the field. These secondary observations serve at the same time as a mechanism for validation of the data on attributes and classifiers gathered, as well as, to some extent, for the ultimate classification results. It will certainly be an advantage if the surveyor is familiar with the land use and management practices in the area.

Provision should be made for registering the specific location of the sampling point within the field (e.g. centre or edge). The adoption of a regular grid for sampling implies that sample points may be taken from anywhere within the plot and possible edge effects need to be accounted for. For that purpose, information on the type of field boundary needs to be recorded as well, knowing that the latter may have a profound effect on the distribution of below-ground biodiversity. While these observations are not of any consequence for the classification of land use, they may provide background information for the analyses of the data (e.g. explanation of possible data outliers). This type of information should be included in the section of data recording sheet that deals with the sampling point registration and administrative data, like the time and date of the recording, the name of the surveyor and so on.

OBSERVATIONS ON LAND USE: CLASSIFIERS AND ATTRIBUTES

Observations regarding the main crop and field size

The first level of the classification system is reserved for observations on the main crop; the main crop being the most dominant vegetation element. The latter means that if the choice is between a tree crop and annual crop, grown in the same field, the tree crop is considered the main crop, as long as the associated ground cover is not classified as sparse; the canopy cover should be classified as 'open' or 'closed'. If the choice is between two or more crops of the same life form, the crop with the highest cover percentage qualifies as the main crop. Provision is made for the description of a second or even third crop.

Classification is based on the life form of the crop. In Table 11.1 the possible (classifier or attribute) values are listed between square brackets. The technical attributes are specific for each of the different life forms, giving a further specification of the life form at that level of detail. In the case of a tree or shrub crop, information can be added on the usage aspect (i.e. purpose) and duration of the crop, because the appearance (vegetation characteristics) may be quite different depending on, for example, the age of a plantation. For herbaceous crops further specification indicates whether the crop belongs to the category of 'root and tuber' crops, 'pulses and vegetables', 'fodder crops' or 'fibre crops'. The most specific information that can be provided is the particular crop type. The categories of herbaceous crop represent an additional level in the classification hierarchy (level of specification) and it is therefore not listed under the heading of 'Technical attribute 1' in Table 11.1. If relevant, one may consider adding a fourth level, providing information on the particular crop variety. Generally the crop type is known. The value domain has not been specified here, but, rather then leaving it open, it is advisable to define a list of permitted names of crop types to prevent any confusion. It is, however, outside the scope of this chapter to provide these lists. For an example, see the list provided by the LCCS (Di Gregorio and Janssen, 2000).

The next level in the classification hierarchy is determined by spatial characteristics of the plot under observation (Table 11.2). The definition of the field size classes 'small', 'medium' and 'large' is not given, as this will depend on the context of the study. It is best to define meaningful classes based on a pilot survey of the range of field sizes present in the area of study (Huising, 1993) with the option to use different definitions

Table 11.1 *Classifiers and technical attributes for the main crop*

Level 1a – Classifier	*Technical attribute 1*	*Technical attribute 2*	
Life form [tree; shrub; herb or graminoids]	***If tree crop or shrub:*** **Leaf phenology** [evergreen; deciduous] **Leaf type** [broadleaved; needle leaved] **Purpose** [nursery stock; harvesting fruits, nuts and other; harvesting of whole plant (e.g. for wood or timber), shade trees] **Duration of the crop** [season (part of the year); 1 year; between 1 and 3 years; between 3 and 10 years; more than 10 years]	**Crop type** (species or variety)	
	If graminoids: **Type** [cereals, bamboos, rice, reeds, grasses]	**Crop type** (species or variety)	
	If non-graminoids: **Category of crops** (I – herbaceous; II – bananas and other tree-like herbaceous plants; III – cover crops; IV – hops and other perennial herbaceous vines)	**Crop type**	
		If herbaceous: **Category of crops** [(root & tubers; pulses & vegetables; fodder crops; fibre crops] (If known specify crop type)	**Crop type**

for each of the various regions in which the study is undertaken, but maintaining the conceptual distinction between small, medium and large. Shape of the field is added mainly as a concern to account for possible edge effects, and further it is an indicator of the degree of organization and mechanization of the farming practices. The 'crop cover' is added here, though it is more related to crop characteristics than to field characteristics. Crop cover is generally used as a parameter for the interpretation of remotely sensed imagery. Here it may serve to indicate the relative importance of crops in multiple crop systems, in cases where the second-level classifier 'crop combination' indicates that there is a second or third crop grown in the field. Further, it provides information on the crop density and is herewith a useful indicator of the intensity of use. In the case of annual crops, the crop cover may be difficult to interpret because of variation in plant architecture, difference in crop growth stages and other factors. Therefore, and especially in the case of annual crops, the density could be directly measured in terms of numbers of plants per unit of measurement (m^2 or ha), or measured in terms of plant spacing (inter-row and intra-row plant distances). Density classes could be specified, but would be very much dependent on the crop type. For permanent life forms, 'closed'

Table 11.2 *Field size and crop-cover characteristics*

Level 1b – Classifier	*Technical attribute 1*	*Technical attribute 2*
Field or plot size [small; medium; large]	**Shape** [square or rectangular; rounded or multi-angle; elongated or strip; irregular] **Field size** (square metres)	**Crop cover** *For non-permanent cover*: [high (>60%); medium (60–30%); low (30–15%)] *For permanent cover*: [closed (>70–60%); open (70–60% to 20–10%); sparse (20–10%)]

cover generally means the crowns interlock. An 'open' cover means that the distance between the perimeters can be up to twice the average crown diameter. For non-permanent covers the class limits are slightly different (see Di Gregorio and Jansen, 2000).

Observations regarding crop combination and cultural practices

Crop combinations are considered at the second level. This refers to whether in one growing season and within the same field, one or several crops are grown. Distinctions can be made according to the number of crops and the sequence of the crops.

In the case of a second and third crop, the spatial arrangement in the field can be specified. The second (and third) crop can be described by the same attributes, at the various levels of specification, as are used for the main crop. The crop type of the second crop is added here as the second technical attribute. A second crop may refer to an understory crop like cardamom, in which case there will not be a specific spatial arrangement. Also in such cases, the total crop cover (i.e. the cover percentages of the individual crops added) may easily reach over 100 per cent. The understory vegetation could also refer to (semi) natural vegetation, as is sometimes the case in plantations. In that case the specification of the life form, perhaps together with leaf phenology and type, would suffice. The second (or third) crop or vegetation element can be incorporated in the class description of the land use system as a descriptive element or predicate (e.g. 'tree crop (rubber) on a large field (plantation), with closed herbaceous understory'). If a particular crop rotation is practised, this should be recorded. Provisions are

Table 11.3 *Crop combination attributes*

Level 2 – Classifier	*Technical attribute 1*	*Technical attribute 2*
Crop combination [single (monoculture); multiple (intercropped)]	***If multiple crop:*** **No of additional crops** [1; 2 or more] **Sequence** [simultaneous, overlapping; sequential]	**(Type of 2nd crop)** ***If simultaneous or overlapping:*** **Spatial arrangement** [1 or 2 row interlaced; alternate strips; fragmented or dispersed]

Table 11.4 *Water supply characteristics*

Level 3a – Classifier	*Technical attribute 1*	*Technical attribute 2*
Water supply [rainfed; post-flooding; irrigated]	***If irrigated:*** **Type of irrigation** [surface; sprinkler; drip]	**Water demand for irrigation** [mm of water supplied per growing season or crop]

made at the third level of classification, dealing with the 'cultivation time factor', to specify that information.

Level 3 in the classification system refers to the cultural practices, defined as water supply characteristics and characteristics in relation to fallowing. The description for the water supply characteristics is straightforward (see Table 11.4).

Post-flooding cultivation is defined, according to the LCCS, as 'when after a field has been flooded with rainwater, the water infiltrated in the soil is used intentionally as a reserve for crop cultivation'. This is similar to, though not the same as, employing water harvesting techniques (addressed separately under the section dealing with land and crop management).

The cultivation time factor relates to the time fraction that the land is used for cultivation of crops. Distinction is made between shifting cultivation, fallow systems and permanent cultivation. Shifting cultivation is defined as land cultivated for less than 33 per cent of the time (Ruthenberg et al, 1980). Fallow systems are defined as land being cultivated between 33 per cent and 66 per cent of the time and permanent cultivation indicates that land is being cultivated for more than 66 per cent of the time. Technical attributes are specified in Table 11.5.

Fallowing refers to the period (growing seasons) during which the land is rested. In principle, distinction should be made between the situation where this is done for the purpose of restoring soil fertility or whether this is due to conditions not allowing the

Table 11.5 *Cultivation time factor*

Level 3b - Classifier	*Technical attribute 1*	*Technical attribute 2*
Cultivation time factor [shifting cultivation; fallowing; permanent]	***If shifting cultivation:*** **Fallowing period** [short term; medium; long term]	**Cropping index** (Ruthenberg's cropping index value)
	If fallowed: **Type of fallow** [bare land; natural; improved]	***If improved:*** **Type of cover crop** (e.g. leguminous, or name type)
	Duration of fallow [short; medium; long]	**Cropping index** (Ruthenberg's cropping index value)
	If permanent: **Permanency** [continuous or intermittent]	***If intermittent:*** **No. of crops in 2 years**

cultivation of a crop (low temperatures or limited water availability). However, for practical purposes and because it has no consequences for considerations of land use intensity, this distinction is not maintained.

Shifting cultivation typically refers to situations where farmers move to open up new plots to farm, whereas in fallowing systems the farmer typically farms the same plot or area of land, of which parts are being rested for varying periods of time to restore soil fertility. Under increased population pressure the fallow period typically tends to be reduced and it is therefore relevant to record the duration of the fallow period.

For shifting cultivation distinction is made between the duration of the fallowing period:

Short-term fallows: having a fallowing period < 1–2 years.
Medium-term fallows: with a fallowing period of > 1–2 years but < 8–10 years.
Long-term fallows: being fallowed for > 8–10 years.

These are partially qualitative assessments and therefore overlapping class boundaries are specified to cater for the specific variation in period of fallowing for any particular area that is probably dependent on the prevailing socio-economic as well as bio-physical conditions. The definition of these class boundaries is based on experience in the field but needs to be confirmed and may be adjusted based on actual data on the length of fallowing periods in the area concerned.

For fallow systems the classification for the period of fallowing is as follows:

Short term: having a fallowing period < 4–5 months.
Medium term: with a fallowing period of > 4–5 months but < 8–9 months.
Long term: being fallowed for > 8–9 months.

The short-term fallows cater for the situation in which there are typically two cropping seasons in a year and the land is rested for the remaining period. The medium-term fallows reflect the situation where the land is rested for six to seven months, which is quite common in certain places – where one crop or two 'short' crops are grown in a year, or where farmers have a longer-term fallow only in the second year. The long-term fallow reflects the typical situation where the land is fallowed for one year or more, but not long enough to classify as shifting cultivation.

Fallows are considered to be 'improved' the moment any planting or seeding is done to change the composition of the fallow vegetation and to improve its quality. No distinction is further made according to the type of crop or the purpose (e.g. for livestock feed). If more precise data are available the Ruthenberg's cropping index can be calculated. The cropping index is given by the following formula (Ruthenberg et al, 1980)

$$CI_r = T_c/(T_c + T_f)$$

where T_c = length of the time of cropping
T_f = fallowing period or length of time the land is not cultivated.

So far, crop rotations have not been considered and they have not been included in the classification system. However, they may a factor to consider in relation to below-ground biodiversity. If a particular crop rotation is practised, a description of the crop sequence is probably best included at the second level in the classification hierarchy where 'crop combination' can be described as 'sequential'. Alternatively provision could be made in Table 11.5 'cultivation time factor', since fallow is often part of a particular crop rotation system. The data should specify the number of years of the crop rotation cycle and the sequence of crops in the cropping cycle.

Land and crop management characteristics

The cultural practices mentioned above include 'water management' as a management operation; land and crop management are considered to represent a separate level in the classification system (Table 11.6a). The major components of the management practices discussed here relate to techniques for land preparation or tillage, the management of weeds, pest and diseases, fertilization and harvesting. 'Land clearing' is added because of its particular relevance at forest margins in the tropics where there is, potentially, a high impact on below-ground biodiversity.

Concerning land preparation and harvesting, the main issue is whether this is done by mechanical means. It is, therefore, an option to combine both classifiers into one that describes the degree of mechanization. Alternatively, the harvesting as a classifier could be excluded from the classification since soil or land preparation is generally the first operation to be mechanized and it has a more profound effect on below-ground biodiversity. Distinction is made based on whether the soil is tilled 'yes or no', and if tilled, based on the type of traction. For animal and mechanical traction, the type of animal or type of machinery employed is indicated at the next level of detail, with reference to the capacity. It is assumed that 'heavy' animals or heavy machinery have bigger impact on the soil, compared to 'light' animals or light machinery, considering both the direct impact of the weight of the animals or machinery and the impact generated by the different types of equipment operated by machinery or animals of different capacity (e.g. depth of ploughing). Further distinction according to actual power requirements or capacity in terms of tractor horse power is not considered relevant because these will depend on local conditions like the type of soil and will not be directly indicative of the degree of soil disturbance.

A technical attribute is added that describes the type of plough used, the options being mouldboard plough, disc plough, chisel plough or ripper. In the order mentioned these ploughs will be, progressively, less efficient at turning over the soil and would leave more of the soil structure intact. Specific mention should be made here if minimum tillage is being practised. Minimum tillage refers to those systems where ploughing and planting is typically combined in one operation, where tillage of the soil is restricted to planting holes, rather then the whole field being tilled, or where the degree of disturbance is minimized through use of specialized equipment, like the ripper/planter common in conservation agriculture (de Freitas, 2000). 'Efficiency' in terms of hours per ha needed for the operation is added as an optional attribute. Efficiency is more of an economic than an environmental concern, in the sense that the higher or lower efficiency will not result in higher or lower impact on the soil and the organisms it harbours; rather the contrary, the soil (and other environmental conditions) will affect the efficiency of

Table 11.6a *Classifiers and attributes related to clearing, tillage and weeding operations*

Level 4 – Classifier	*Technical attribute 1*	*Technical attribute 2*
Clearing [no clearing; manual; mechanical; chemical]	***If manual:*** **Mode** [slash and burn; slash and mulch] ***If mechanical:*** **Grade** [light clearing; intermediate clearing; heavy clearing]	
Tillage [no tillage; manual (hoe); animal traction; mechanical]	***If manual:***	**Extent of tillage operation** [complete; reduced or minimum tillage] **Efficiency** (hours/ha)
	If animal traction: **Type (and number) of animal used** [buffalo; bullock; cow; horse; mule/donkey]	**Type of plough** [mouldboard, chisel; minimum tillage/ripper-planter] **Efficiency** (hours/ha)
	If mechanical: **Class of machinery** [two wheel (light); four wheel (heavy)]	**Type of plough** [mouldboard; disc; chisel; minimum tillage/ripper-planter] **Efficiency** (hours/ha)
Weeding [no weeding, hand weeding; mechanical; chemical weed control]	***If hand weeding:*** **Type of weeding** [pulling; hoeing]	**Efficiency** (hours/ha)
	If mechanical: **Type of weeding** [frequent ploughing; cultivation for weed control]	**Type of operation or equipment** [e.g. push hoe; rotary hoe] **Efficiency** (hours/ha)
	If chemical: **Type of equipment** [rucksack sprayer; other mechanized]	**Application volume (Roundup)** [LV (< 850g ae/ha); IV (between 850 and 2500g ae/ha); HV (> 2500g ae/ha)] **Herbicide, active ingredient and application rate**

the operation. It is included here for compatibility with methods that incorporate fossil fuel use in their calculation of land use intensity. Fuel use can be easily estimated if efficiency, type of equipment and type of traction is known. Optionally, fuel consumption could be added as an attribute at the next higher level of specification.

The main concern related to the control of weeds, pest and diseases is the use of agrochemicals. Therefore, one could consider combining both into one classifier that describes the use of agrochemicals (inorganic fertilizers excluded). It is considered, however, more straightforward and practical to separate them. The classifier for weed control specifies the mode of control mechanisms, whether this is herbicidal, mechanical, biological or cultural. In case of cultural (and biological) control measures (for example crop rotation, use of fallow, etc.), these will have been addressed by the recording of land use practices in the previous levels of the classification system, and it is therefore not included in the value domain.

'No-weeding' may cater for those situations where no actual removal of the weeds occurs. With respect to hand weeding, distinction is made between whether this is done by pulling or using equipment like a hoe. With respect to the use of chemicals for weed control, distinction is made according to how it is applied, though use of a rucksack sprayer will be the most common. Further differentiation based on volume of application is not feasible considering the vast number of herbicides that are available on the market, the different types of active ingredients and the various formulations that exist (Oregon State University, 2005). However, given that Roundup is, by far, the most used herbicide in the tropics (www.weeds.iastate.edu/mgmt/2004) classes for the amount of its application are provided, specified in grams of acid equivalents (ae) per hectare of glyphosate, which is the active substance in Roundup. The classes are somewhat arbitrarily assigned, based on general recommendations for application on the various crops and converting from pounds per acre to grams per ha (weeds.ippc.orst.edu/pnw/weeds). An application rate of 'ae' less then 850g ha^{-1} is considered a low volume (LV) of application, whereas application of 'ae' of more than 2500g ha^{-1} is considered a high volume (HV). If more detailed information is required, for Roundup or other types of herbicide, the active ingredient (ai) should be specified together with the amount (grams) of the active ingredient per hectare. If considered critical, it is recommended to devise a classification system for the application volume of herbicides that is relevant for the area being surveyed. The classes for weed management do not cover all possible measures to control weeds and therefore the recording sheet should provide an additional entry for observations regarding any specific weed control measures.

With respect to the measures for the control of pests and diseases, similar classes are defined; namely, natural control or cultural measures, biological control or chemical control mechanisms (Table 11.6b). Concerning the use of pesticides, fungicides or other chemicals, distinction is made based on whether they are broadly applied to the whole field or whether the application is restricted to particular spots, and based on the type of equipment. The volume classes for spraying of pesticides are obtained from Craig et al (2002) and apply to aerial spraying. It is assumed that the mode of application will not be of much consequence for the application rate, and the same volume classes are considered relevant, irrespective of the mode of application. The volumes specified relate to the liquid form in which the pesticide or fungicide is obtained, without considering the further dilutions that are made before actual application. Again, if relevant, actual amounts applied could be specified.

For both inorganic and organic fertilizers, classes for the application rate are provided. For manure, these are based on the assumption of a 30 per cent dry matter content. Generally 5–10 tonnes/ha are assumed to be required to maintain humus levels of the soil; these quantities being supplied once every four to six years (ILACO, 1985). The frequency of application, however, may vary and in the table below the figures are therefore converted to yearly application rates. Low application rates of less than one tonne per ha per year are considered insufficient to maintain humus and organic matter levels in the soil. Two tonnes per ha per year are considered sufficient for all different crop types.

With respect to application of inorganic fertilizers, the ranges of application rates are specified for both lower and upper limits of the classes, rather than one particular value. This is done to accommodate for the different grades of fertilizers. Lower-grade fertilizers have less than 25 per cent plant nutrients, whereas high-grade fertilizers contain up to 50 per cent nutritive elements. Where lower grades of fertilizers are used, the higher values in the specified ranges should be assumed. The nutritive elements may refer to one single element (e.g. N) in straight fertilizers or the combined elements in incomplete mixtures or complete mixtures. (The weight percentage relates to N for nitrogen, P_2O_5 for phosphorus and K_2O for potassium). The rating takes 75kg ha^{-1} of N as a reference point for the application of N. This would be the recommended application for a moderately fertile soil in relation to an expected yield of around four tonnes of maize (grain), for about three tonnes of wheat on reasonably fertile soils and 25 tonnes of cassava on a moderately fertile soil. An application of 75kg ha^{-1} of N would require 150kg ha^{-1} of high-grade N fertilizer and around 350kg ha^{-1} of low-grade fertilizer. These values are therefore taken to be the upper limit of the class of 'medium fertilizer application rate'. The other classes are derived from these figures, with a very low application rate representing around 10 to 15 per cent of this reference amount. If application rates are generally high, or if very high application rates are of particular concern, one might consider adding a class for the very high application rates that would relate to applications of 300kg ha^{-1} or more of a high-grade fertilizer and of 700kg or more per ha of a low-grade fertilizer. The placement of the fertilizer, whether in or around the planting hole, will only be relevant in case of low or very low application rates.

The way in which a plot of land is cleared may have a profound impact on belowground biodiversity and it is therefore included as a separate classifier. This will be specifically relevant where shifting cultivation is practised or where forest is or has been converted into agricultural land in the recent past, which is the case in some forest margins where the CSM-BGBD project operates. Clearing is also relevant in the case of fallow systems, especially when it involves burning.

'No clearing' (see Table 11.6a) is listed when the land has been cleared from forest or secondary vegetation more than 20 years ago, which coincides more or less with the maximum duration of fallow under shifting cultivation systems. The clearing may include operations for levelling of the land. Further distinction is made based on the type of equipment used. 'Light clearing' is indicated when both manual and mechanical means are used, where trees are felled by hand or using chain saws and the trunks are removed using tractors or other relatively light machinery. 'Intermediate clearing' signifies that the land is being cleared completely with the use of machinery, using power saws and shearing blades for cutting vegetation, and bulldozer blades, anchor chains and other equipment for clearing of vegetation. 'Heavy clearing' makes use of crawler tractors and other heavy machinery. In all the systems it is important to record whether burning takes place to get rid of the dead vegetation or whether this remains in the field as mulch. Currently distinction is made between 'Slash and Burn' and 'Slash and Mulch' systems. However, any incidence of fire as part of the land management practices should be recorded under this section as an important feature.

Table 11.6b *Classifiers and attributes for pest and disease management, fertilizing and harvesting*

Level 4 – Classifier	*Technical attribute 1*	*Technical attribute 2*
Pest and disease management [preventive measures and natural control; chemical control; biological control]	***If chemical:*** **Extent of spraying** [spot spraying, field spraying]	**Volume of application** [ULV (<10L/ha); LV (10–100L/ha); IV (100– 200L/ha); HV (>200L/ha)]
	Equipment used [manually operated backpack or handheld sprayer; engine sprayers; ground vehicles; aircraft]	**Chemical type and application rate**
Fertilizing [no fertilizers; manures; chemical fertilizers; combination of chemical and organic fertilizer]	***If organic fertilizers:*** **Type of fertilizers** [crop residue; green manure; compost; FYM] **Placement** [pits or pockets; strip; broadcast]	**Application rate** [low (<1t/ha/yr); medium (1–2 t/ha/yr); high (>2 t/ha/yr)]
	If inorganic: **Class of fertilizer** [straight; incomplete mixtures; complete mixtures]	**Application rate** [very low (<20– 60kg/ha); low (between (25–60) and (75–175)kg/ha); medium (between (75–175) and (150–350)kg/ha; high (>150–350kg/ha)]
	Placement [sowing pits or pockets; strip; broadcast]	
Harvesting [manual; mechanical]		**Efficiency** (hours/ha)

Other observations on land and crop management

In addition to registering the characteristics of land and crop management as described in the tables above, it is recommended that observations on practices for soil and water conservation and on the status of the crop should also be included. Soil and water conservation measures will effect populations of soil organisms indirectly by improved water storage, increased retention capacity and reduced soil loss and will therefore present relevant information. Information on conservation measures will not be captured directly by the recording of the attributes listed above. The recording sheet should therefore provide for the description of soil and water conservation measures, including those related to water harvesting techniques.

The 'World Overview of Conservation Approaches and Technologies' (WOCAT), provides a good overview on their website of the many techniques that exist (www.wocat.net). Probably the best way of categorizing them is by whether they involve agronomic measures (like mixed cropping, contour cultivation, mulching, etc.), use of vegetation elements (referring to grass strips, hedge barriers, windbreaks and other) or structural measures that include terraces, banks, bunds, constructions and palisades among others. Water harvesting techniques like a retention ditch, pitting and other are then included as structural measures. If the technique is know by a particular name (for example 'Zaï' that is practised in Niger) this information should be provided.

Observations on the status of the crop could be useful to corroborate earlier findings (or classification results) especially when related to management. Information on crop yield would be useful in the same manner, but not as a basis for classification. Crop yield as an indictor integrates many different factors and it is difficult to obtain reliable and correct figures. Field-based yield assessment could include the following: plant population per square metre, number of tillers on individual cereal plants, relative height of the growing crop and relative diameter of growing crop (see Stocking and Murnaghan, 2001). These provide for additional information on the main crop as specified at the first level of the classification system. Alternatively, crop growth characteristics can be used as a proxy for yield. These may in general be described as one of the following: stunted growth with height of the crop clearly lagging behind compared to a vigorously growing crop; low plant vigour with diameter of the crop lagging behind and general discoloration of the leaves and crop; vigorously growing crop. There are guides available for the registration of specific nutrient deficiencies in crops (see Stocking and Murnaghan, 2001, for signs of nutrient deficiencies in maize, beans and cabbage).

Observations on crop status should be accompanied by observations on the incidence of crop pests and diseases that should include percentage of the crop affected and the severity of the infection in terms of degree to which the plants are affected. If the specific pest or disease is known, the name should be listed. It is outside the scope of this chapter to provide more detailed information on pests and diseases and their classifications.

Field distribution patterns and trees on farm

Observations concerning the type of fields and permanent woody vegetation elements within the plot under observation and its direct surroundings are included. This allows for the inclusion of woody vegetation elements in the description that are not captured with the system for description of land use so far. This is the case when tree cover is sparse (less than 10–20 per cent) which may occur with trees scattered within the plot, living fences and other important vegetation elements, both at plot and landscape level. Further it allows assessment of the land use at the particular sampling point within the context of the land use of the surrounding area (i.e. it helps to establish whether the land use at the plot under observation is representative or not), which is important for the mapping of land use, and finally it allows for validation, and to some extent verification, of the observations at the sampling location. The land use in the directly surrounding area of the sampling point will have influence on the below-ground biodiversity component of the plot under observation. For example, in case of shifting cultivation or fallow systems, one would expect to find fallows to be part of the land use pattern; the Ruthenberg factor in principle translates into a spatial ratio between fallow and cultivated land in current land use. Fragmentation of land use that can be assessed from these observations is in itself an important aspect of land use intensity.

Table 11.7 *Field and land use distribution characteristics*

Spatial classifier	*Technical attribute 1*	*Technical attribute 2*
Spatial distribution of fields [continuous; clustered; scattered]	***If continuous:***	
	Field size distribution [even; uneven]	***If even:*** **Size class** of (majority of fields) [small; medium; large]
	Field pattern [regular; irregular]	**Main crops** (list crops of four neighbouring fields)
	If non-continuous:	
	Part of cultivated land (% of cultivated and managed area)	**Main crops** (list crops of four closest fields)
	Part natural vegetation cover (% of (semi) natural terrestrial area)	**Land cover** (give dominant land cover of non-cultivated terrestrial area)
	Part cultivated aquatic area (% of cultivated aquatic area)	**Type of crop or activity**
	Part non-cultivated aquatic area (% of non-cultivated aquatic area)	**Type of aquatic area**
	Built-up area (% terrestrial artificial surface)	**Type of built-up area** (e.g. residential; industrial; etc.)
	Bare soil and land (% bare areas)	
	Water bodies (% water bodies)	**Type of water body** [artificial; natural] (specify type, like pond; reservoir; swamp, if possible)]

With respect to the distribution of fields (assessment of field distribution pattern), the first distinction is made between the fields being continuous, distributed in clusters or scattered (see Table 11.7). Secondly, the size, shape and pattern of the fields are described. And finally, there is an option to specify the percentages of the area under the various major land use categories: terrestrial cultivated and managed areas, semi-natural terrestrial areas, aquatic or regularly flooded cultivated areas, (semi) natural aquatic vegetations, artificial surfaces, bare areas, or artificial water bodies and natural water bodies (snow and ice), using the major land-cover classes of the LCCS. The various land use components can be described in more detail, which seems relevant especially in the case of the terrestrial cultivated areas, by specifying the main crops, water supply characteristics and permanency of the cultivation. The field pattern relates to the shape of the field and spatial arrangement of the fields.

The last element that needs to be recorded is the presence of permanent elements of woody vegetation in the landscape. We will refer to these as trees on farm (TROF). It is assumed that woodlots and other plots of woody vegetation are covered by the method for land use classification as described so far. For low tree-cover percentages and trees scattered within the field or aligned as living fences, this might not be the case, though these elements may be a characteristic feature of the land use and land cover and may be a major influence on the presence of particular soil biota. The first attribute describes the

Table 11.8 *Characteristics of trees on farm*

Level 5 – Classifier	*Technical attribute 1*	*Technical attribute 2*
Type of TROF [no TROF; living fences; trees scattered within field; wind breaks or other linear woody elements; parklands]	***If living fence:*** **Tree species**	**Coverage** (give percentage of the field boundaries that are constituted of living fences)
	If scattered within field: **Purpose** [mainly fruit and nuts; mainly wood and timber; mainly shade trees; other]	**Density** (no of trees per ha)
	If wind breaks or other linear arrangements: **Type or species**	**Total length per ha** (m/ha)
	If palisades: **Dominant type or species**	

spatial distribution characteristics of the trees. 'Parklands' refers to trees spaced in rows as in palisades. Table 11.8 specifies the attributes used to describe the TROF. The technical attributes allow for the description of the type of (the most dominant) tree (either by species or other taxonomic level) and the purpose for having the trees (e.g. harvesting of fruits and nuts, for wood or timber, for fuel wood or building materials). Further information concerns the coverage or density of the tree component, specified in terms of number of trees per ha or length of the linear woody vegetation elements. Information on trees belonging to the family of leguminous trees is useful in relation to the occurrence of legume-nodulating bacteria. As already mentioned and in view of the above it is important to register the location of the sampling point with respect to the field boundary.

DESCRIPTION AND CLASSIFICATION OF LAND USE

Definition and description of land use classes

It is not the purpose of this chapter to describe in detail how land use classes are defined. It is important to derive classes in a systematic way from the classifiers and attributes presented here, using decisions rules, such that the one set of classes defined for one particular area can be easily mapped onto the set of classes used for another area and vice versa. The principles of the definition of land use classes, the design of a classification hierarchy and the attribution of class values to the observations in the field are explained in this section.

Classes are defined by the particular combination of classifiers and attributes (determined by the value the classifier or attribute takes). Not all attributes or classifiers need to be considered and this determines the level of detail at which the classes are defined. If only the classifiers of the first three levels in the hierarchy are considered, a land use class could be defined by the following attribute values: 'tree crop', 'large plot size', 'single crop', 'rainfed' and 'permanent cultivation'. This would translate into a 'tree plantation'. If additional attributes (e.g. crop type) of the main crop are considered as well, it would be possible to distinguish, for example, between teak plantation and rubber plantation. On the other hand, additional technical attributes, when they are considered, do not always have to imply the definition of an additional set of classes at the next level in the classification hierarchy. For example, different degrees of tree cover may be observed within the 'tree plantation' class (or even within the teak plantations, for example) but this may be considered not to be relevant for defining separate classes of tree plantations. Rather the information on cover percentage is retained as a particular attribute value for each individual observation (instance or occurrence) of the object class (i.e. tree plantation).

A classification hierarchy is established in a similar way, by deciding on the precedence of one classifier over the other. This refers to the levels in the classification hierarchy as presented above and not to the levels of generalization (the second axis in the classification system). Because the system as presented deals with cultivated and managed areas, the hierarchy has the main crop as main entry point. However, in areas where shifting and permanent cultivation are practised side by side, the 'cultivation time factor' might be given precedence over the main crop as the dominant characteristic. However, this feature is generally characteristic for larger areas, rather than for plots, and should therefore be used for smaller-scale mapping exercises, in which case the 'cultivation time factor' could be used as a parameter in aggregation hierarchy rather than in the classification hierarchy for plot observations. For explanation of classification and aggregation hierarchies see Huising (1993).

Table 11.9 provides a generalized list of the classifiers and class attributes arranged according to the levels of the classification hierarchy. Because of the nested structure not all attributes can be easily arranged in one single table. For the precise information refer to Tables 11.1 to 11.8. Table 11.9 provides an overview of the principal classification structure. Classes are assigned by decision rules, and herewith the classes are defined at the same time. The decision rules generally take the form of a set of IF – THEN statements that apply to the attributes, combined through Boolean operators (i.e. AND, OR, XOR and NOT). For example:

> IF (<life form> equals 'tree' and <field size> equals 'large' and <crop combination> equals 'single' and <cultivation time factor> equals 'permanent') THEN <class> = 'tree plantation'.

To illustrate the class definition and description, consider the values specified for the various classifiers and technical attributes as given in Table 11.10 below. The classifiers specify a 'tree crop', a 'large field', 'no irrigation' and a cultivation being 'permanent', meaning a large tree plantation. Considering the information on main crop (level 1 classifiers), the plot would be described as a large (with 10,000m^2 for plot size), more than ten-year-old teak plantation (forming a closed canopy), with 100 trees per ha, and receiving 1200mm of rain per year. Using the information on tree species allows us to distinguish between other tree plantations for timber or wood production and using the information on purpose allows us to distinguish these plantations from orchards, cacao plantations and other.

Some of the land and crop management classifiers and attributes are (often) not relevant for plantation crops, because field preparation, pest and disease management

Table 11.9 *Level 1 to 5 classifiers and attributes for the classification of land use*

	Level 1 Crop and field characteristics				*Level 2 Crop combination*			*Level 3 Cultural practices*		
Classifier	Life form main crop	Field size class		Crop cover	Single or multiple crops			Water supply	Cultivation time factor	
Attributes Level of spec 1	Type of crop (category)	Field shape	Field size	Crop cover class	No of crops	Time sequence	Spatial arrangement	Type of irrigation	Length of fallow period	Type of fallow
Attributes Level of spec 2	Specific crop (cat. or species)			Plant density or spacing	Type 2nd crop			Amount of water supply	Ruthenberg index	Specific type of fallow
Attributes Level of spec 3	Crop variety (species or var.)				Type 3rd crop					

	Level 4 Management						
Classifier	Means of clearing		Field preparation mode	Mode of weeding	Pest & disease management	Fertilizing practices	Mode of harvesting
Attributes Level of spec 1	Grade of clearing	Mode of clearing	Type of traction	Type of operation / Frequency	Extent of operation / Frequency	Type of fertilizer	
Attributes Level of spec 2			Type of plough / Extent of operation	Type of equipment	Type of	Placement, mode of equipment	application
Attributes Level of spec 3			Efficiency	Application rating	Application rating	Application rating	Efficiency
Attributes Level of spec 4				Herbicide spec / appl volume	Chemical spec / appl volume	Fertilizer specs / appl volume	

	Level 5 Spatial configuration of fields and trees on farm					
Classifier	Spatial configuration of trees on farm		Distribution pattern of fields			
Attributes level 1	Purpose of trees on farm	Tree coverage (crown)	Part cultivated and managed terrestrial area			Part non-cultivated terrestrial
Attributes level 2	Type of dominant tree species	Density of trees or length of tree lines	Field size distribution pattern	Field spatial arrangement	Life form main crops	Type of other use or cover
Attributes level 3	Tree species		Size or size class of field		Type main crops	% coverage

Table 11.10 *Example of possible classifier and attribute values of a large teak plantation*

	Level 1 *Crop and field characteristics*					*Level 2* *Crop combination*			*Level 3* *Cultural practices*		
Classifier	Tree		Large			Single			Rainfed	Permanent	
Technical Attributes Level 1	Evergreen, broadleaved Wood	>10 yrs	Regular	10,000m^2	Closed	Natural vegetation*	n/a	n/a	n/a	n/a	n/a
Tech Attributes Level 2	Teak				100 trees/ha	n/a			1200mm**	n/a	n/a
Tech Attributes Level 3	*Tectona grandis*					Na					

Note: * In this case it is possible to describe the undergrowth in the plantation if relevant. This undergrowth would belong to the category of (semi) natural vegetation for which other parameters are used to describe it, though not discussed in this chapter.

** In case of rainfed systems it is possible to specify the amount of rainfall, if one has specific details, though this is generally described under site characteristics

Table 11.11 *Example of classifier and attribute values for smallholder maize plot*

	Level 1 Crop and field characteristics				*Level 2 Crop combination*			*Level 3 Cultural practices*		
Classifier	Graminoids	Small			Multiple			Rainfed	Permanent	
Attributes Level of spec 1	Cereals	Square	2000m^2	Medium to High	1 additional crop	Simultaneous	2 rows alternate	n/a	Intermittent	Improved
Attributes Level of spec 2	Maize			75 x 25cm	beans			n/a	0.16	Mucuna
Attributes Level of spec 3										

Level 4 Management

n/a	Animal traction	Mechanical and chemical	None		Chemical fertilizer		Manual
n/a	2 oxen	Frequent (2x) ploughing and spraying once	n/a	n/a	Straight	Pockets	
n/a	Mouldboard plough	Rucksack sprayer	n/a		VL (Very low)		
	na	LV (Roundup)	n/a		50kg CAN/ha		

Level 5 Spatial configuration of fields and trees on farm

No TROF		Continuous			
n/a	n/a	90%			10%
n/a	n/a	Even	Regular	Herb and graminoids	Roads – infrastructure
n/a		Small		Peas, beans, maize	n/a

and fertilization generally do not take place. Weed control measures may be taken in the form of slashing (and mulching) or indeed by the use of chemicals. The management of these plantations therefore generally will classify as 'extensive'. Distinction in land use intensity between the various plantations could be made based on the level 1 class attribute 'tree density', or the level 2 attribute crop combination of the trees, which would only make sense if there is a lot of variation in these characteristics within the area and if it would reflect different management regimes. This would, for example, allow for the distinction between 'jungle' rubber, a rubber-based agroforestry system in which rubber trees (*Hevea brasiliensis*) are grown within the forest (Joshi et al, 2002), and pure stands of rubber trees.

As for the level 5 classifiers, information on surrounding land use is important to establish, for example, whether the tree plantation under consideration is a small isolated woodlot in an environment dominated by annual crops or part of a landscape dominated by plantations of tree crops.

Table 11.11 provides the data for an imaginary plot of maize. The first level specifies that it concerns a small plot with maize as the main crop, the second level indicates that the maize is intercropped and the third level that the field is permanently cultivated with a short improved fallow once in two years. The main land use category to which this particular plot would belong could be defined as 'small-scale permanent cultivation with maize as the main crop'. A sub-class could then be defined adding the description 'intercropped with a legume crop' to distinguish it from maize as a pure stand. Alternatively, if the information is accurate and considered important in the context of the classification, the criteria for improved fallows could be considered for defining sub-classes. Specific information will relate to the plant density and specific arrangement of the crop combination and to what species is used for the improved fallow, without this being reflected in the definition of the classes.

Management and input level: the information on crop and land management tells that the level of soil disturbance is intermediate (the plot having been ploughed twice using animal traction) and that the chemical input has been low, with a low volume of herbicides being used, no fungicides or pesticides applied and only a very low amount of fertilizer applied. This could translate, depending on the classification rules applied, into 'with management characterized by use of animal traction and low inputs of agrochemicals'.

Level 5 type classifiers and attributes inform us that the land is continuously cultivated with mainly annual crops, that the fields are generally small, with no or little presence of trees, indicating rather fragmented land use patterns and intense use of the land.

This section intends to demonstrate how a set of classes could be defined and organized in a hierarchical structure and that the system is rather flexible in the choice of classifiers and attributes to be considered for the classification. A standard classification scheme would not work, because the definition of the set of classes depends on the context and specific purpose of the classification.

Ordering of land use classes with respect to intensity of use

In agricultural systems, management aims to modify or set conditions that are conducive to the production of a crop. The degree of interference with (disturbance or alteration of) the natural ecosystem would therefore be a good measure of land use intensity. The general notion is that cultural and managed systems are more intensive than natural, non-managed systems, with overall intensity increasing with the intensity of cultivation and the intensity of the management (the more management input required to maintain the systems, the more intense the use).

There seems to be general agreement on the factors that determine land use intensity, but definition of a land use intensity index or measure of land use intensity is not straightforward. Giller et al (1997) defined a land use intensity index that considers frequency of soil occupation (as expressed by the Ruthenberg index), nutrient use, pest management, energy input and water management as relevant factors. There have been several attempts to define a measure of land use intensity, but these have not been particularly successful, as it is difficult to assign proper weights to the explanatory variables, reflecting the degree to which these factors have an effect on the ecosystems (the degree to which they alter or disturb the natural ecosystem) and because of the sometimes prohibitive data requirements for a quantitative assessment of land use intensity. Furthermore, efforts to devise a universal index or measure of land use intensity will be fruitless, because every index is a particular expression of land use intensity, with the relevance of that particular index depending on the context or purpose for which the index is being developed.

For the purpose of investigating trends in the loss of below-ground biodiversity in relation to an increase in land use intensity, ranking of land use classes in terms of land use intensity (or alternatively a definition of land use classes that reflects different levels of land use intensity) is required. This is done by considering those attributes (and classifiers) that express a particular aspect of land use intensity. There are two principal axes along which land use intensity is measured. One relates to the permanency (or frequency) of operations and the second to the intensity (or amplitude) of the operations. In both cases the cover of the (semi) permanent vegetation component is a useful indicator. Namely, in case of shifting cultivation, the proportion of time the land is covered by semi-permanent secondary regrowth directly translates into the frequency of cultivation. The presence of permanent vegetation in the cropping system will limit possibilities for tilling the soil or the proportion of the land that can be tilled. Also, though not directly related to the influence of the management operation, the presence of permanent vegetation will reduce the influence of climatic variation on the soil ecosystem, indicating that the system with the larger proportion of permanent vegetation (either as natural or planted vegetation elements) represents the lower intensity of use, or lesser degree of disturbance.

This allows arrangement of vegetative components (objects) along two gradients, much in the same way as illustrated by Kuechler and Zonneveld (1988). The first gradient represents the presence (proportion in space or time) of semi-natural vegetation, which in the forest margin environment translates into woody vegetation elements. The second gradient represents the cover of woody vegetation within the cropping system. The objects relate to naturally occurring trees (in the case of natural vegetation elements), to planted or cultivated trees (viz. trees crops) or to cultivated non-woody vegetation elements (e.g. annual crops and pasture) in the case of cultivated vegetation elements. The continuum can be graphically represented by a triangle in which the three corners represent the extreme values of the land cover consisting either of only naturally occurring trees (i.e. forest), of only planted trees (plantation) or of only annual crops or

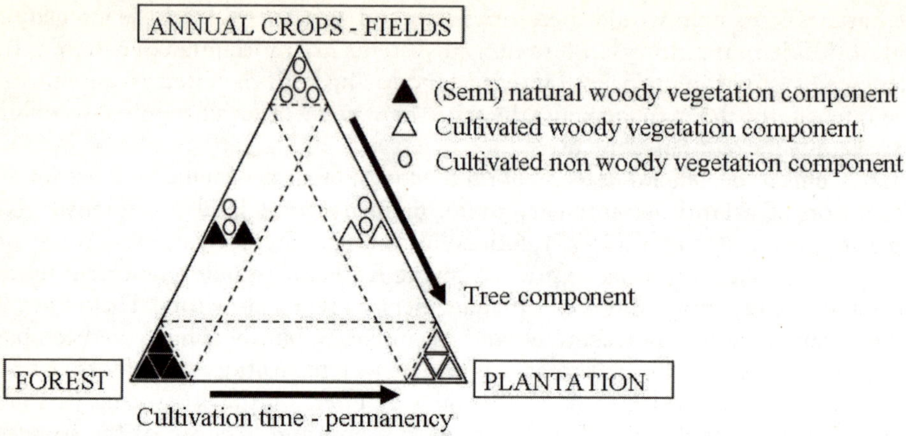

Figure 11.1 *Arrangement of land use and cover classes based on presence and absence of natural vegetation and presence or absence of a tree component*

pasture. Based on classifier and attribute values of the classification system the location along these two axes is determined. First, the permanency is expressed by the 'cultivation time' factor. The classification system allows for arranging the land use classes in the order of increasing intensity: 'no cultivation' – 'shifting cultivation' – 'fallowing systems' – 'permanent cultivation' (see Table 11.5). A further subdivision can be made based on the duration of the fallowing period as specified by technical attribute 1, at the next level of specification. Besides a partitioning in time, this axis likewise expresses a partitioning in space between natural and cultivated vegetation elements, which is relevant for particular land use systems like the 'jungle rubber' system mentioned earlier.

Secondly, within each of the classes thus defined, but most relevantly for the system of permanent cultivation, the land use objects are ordered along the second gradient of land use intensity, according to the tree and shrub component in the cropping system (again referring to proportions either in space or in time). Ordering of the classes according to this gradient is done using the main crop characteristics (Table 11.1) together with crop-cover characteristics (i.e. percentage of crop cover, Table 11.2) and crop combinations (Table 11.3). If the tree cover is sparse, information related to trees on farm (Table 11.5) has to be considered.

If solely the life form of the main crop is considered, tree crops are considered less intensive compared to shrub crops and to crops of graminoid and herbaceous life forms. In the case of tree crops (tree-based systems) we can consider the number of strata of the vegetation and soil cover percentage for further ranking. For example, tree plantations are considered less intensive than 'shrub-like' tree crops like coffee and tea. Further differentiation for non-tree crops is based on the permanency or crop duration, as reflected by the information on crop category (see technical attribute 1, Table 11.1). With respect to the type of graminoid crop, the 'bamboo' class comes before 'grasses' and the 'cereals' and 'rice' classes (see Table 11.1). With respect to the non-graminoid crops, the order of intensity of use would be as follows: 'banana and other tree-like herbaceous plants', 'hops and other perennial herbaceous vines', 'cover crop' and 'herbaceous crop'.

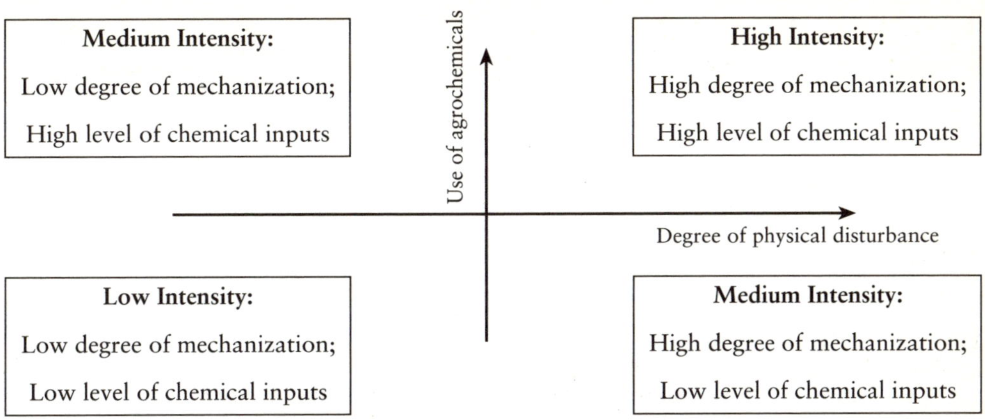

Figure 11.2 *The four quadrants of intensity defined by level of mechanizations and use of agrochemicals*

In the next step, the intensity of the operations has to be considered. Again, within each of the classes resulting from the two steps explained above, further ranking is carried out based on management characteristics (the fourth level in the classification system). There are two aspects to be considered in relation to management: one relates to the degree of the physical disturbance inflicted by management operations and the other relates to the degree of chemical interference with the system (use of agrochemicals). Level 4 classifiers and attributes (especially in relation to tillage operations, weeding operations and pest and disease management) are used to define classes of intensity. At the more general level, four classes of intensity are defined considering the degree of mechanization (i.e. tillage operation) and the degree of use of agrochemicals (for weed control and pest and disease management) as indicated in Figure 11.2. Allocation of the class of intensity is based on the value for the classifier for tillage operations (i.e. no till, manual, animal traction and mechanical means). One could also consider the weeding operations here, with 'mechanical weeding' further adding to the level of intensity.

Further differentiation can be made based on the type of animals or machinery used (ranked according to horsepower provided) and the type of plough, with the mouldboard plough having the most profound effect on the soil (complete turnover), followed by the disc plough, the chisel plough and lastly by the ripper-planter, representing the lowest level of disturbance.

For the ranking order with respect to the use of agrochemicals, the classifiers for weed control and for pest and disease management are considered in conjunction. At the most general level, only the 'use' or 'no use' of chemicals need be considered, but further differentiation can be made based on application volume (as reflected by the placement being selective or localized, type of equipment or machinery used in case of pest and disease management, and volume of application, see Table 11.6b).

The hierarchical procedure described above will result in clearly separated branches in the classification tree and fails to recognize that there may be a considerable overlap in terms of land use intensity between the classes. This is illustrated in Figure 11.3, with

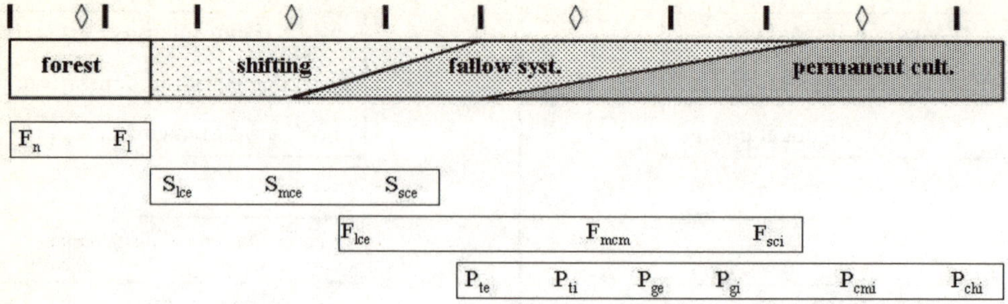

Where		
	Fn	- Natural Forest
	Fl	- Logged Forest
	Slce	- Extensive crop, under shifting cultivation (fallowing period > 8-10 yrs)
	Smce	- Extensive crop, under shifting cultivation (1-2 yrs < fallowing < 8-10 yrs)
	Ssce	- Extensive crop, shifting cultivation (fallowing < 1–2 yrs)
	Flce	- Extensive crop, fallow system (fallowing > 8-9 mths)
	Fmcm	- Medium intensity crop, fallow system (4-5 mths < fallowing < 8-9 mths)
	Fsci	- Intensive crop, fallow system (fallowing 4-5 mths)
	Pte	- Extensive tree crop, permanent cultivation
	Pti	- Intensive tree crop, permanent cultivation
	Pge	- Grassland, extensive, permanent cultivation
	Pgi	- Pasture, intensive, permanent cultivation
	Pcmi	- Medium intensity crop, permanent cultivation
	Pchi	- Intensive crop, permanent cultivation

Figure 11.3 *Ranking of land use in terms of land use intensity*

exemplary classes. The overlap will be considerable between the classes of the fallow systems and the permanent cultivation systems, at the first step of the classification, considering that a tree plantation (viz. permanent cultivation under extensive management) will represent a far less intensive system than an intensively managed crop under a short fallow system.

In order to arrive at a final ranking order of land use classes in terms of land use intensity, these classes have to be positioned relatively to each other, as illustrated above. In the end a pair-wise evaluation of the individual land use classes in the region of overlap will give the final ranking of the classes in terms of land use intensity. As illustrated in Figure 11.3, the most intensive land use class belonging to 'shifting cultivation' is followed by the least intensive land use under 'permanent cultivation' in terms of intensity of use, considering that the tree cover in tree plantations may amount to some 60 to 70 per cent, which more or less corresponds to the 66 per cent of the time that the land is not cultivated, and that is taken to be the upper limit for the class of 'shifting cultivation'.

For comparison of the individual land use classes, in the region of overlap between the major land use categories, the same criteria apply that are used in the second and third step of the classification, which determine the order along the second land use intensity gradient, described above. That is, in the first instance the life form of the dominant vegetation or crop is considered, not making any distinction, at this stage, whether this dominant vegetation is either from cultural or (semi) natural origins. (For example, if the land is fallowed for 60 per cent of the time and consists of grass vegetation, the grass vegetation is considered the dominant vegetation type.) In the second instance the intensity of the management is considered (in relation to the most demanding crop in the system, that is, in case of the 'fallow systems', referring to the cropping component rather than to the fallow) to allow the class under consideration to jump one or two positions in the ranking order.

In this way, permanent grassland, even if intensively managed (P_{gi}) will rank lower in terms of land use intensity compared to an annual crop under a short fallow system (F_{sc}). Likewise an intensively managed crop under a short fallow system (F_{sci}) may jump a position in comparison with a permanently cropped area under low- or medium-intensity management (P_{cmi}). With respect to forest, distinction is made between natural forest, managed forest and logged forest according to the land use systems in the humid tropics as distinguished in the ASB project (Bignell et al, 2005), introducing the notion that extraction of (forest) products serves as an indicator for intensity of use.

Acknowledgement

We are grateful to Professor Ken Giller for comments and to Dr Peter Okoth for the discussions in preparation of the chapter.

References

Bignell, D. E., Tondoh, J., Dibog, L., Huang, S. P., Moreira, F., Nwaga, D., Pashanasi, B., Susilo, F.-X. and Swift, M. (2005) 'Belowground diversity assessment: Developing a key functional group approach in best-bet alternatives to Slash-and-Burn', in C. A. Palm, S. A. Vosti, P. A. Sanchez and P. J. Ericksen (eds) *Slash-and-Burn Agriculture: The Search for Alternatives*, Columbia University Press, New York

Craig, I. P., Woods, N. and Dorr, G. J. (2002) 'Aerial application', in D. Pimentel (ed) *Encyclopaedia of Pest Management*, Marcel Dekker Inc, New York

Di Gregorio, A. and Jansen, L. J. M. (2000) *Land Cover Classification System (LCCS): Classification Concepts and User Manual*, GCP/RA/287/ITA Africover – East Africa Project and Soil Resources, Management and Conservation Service, FAO, Rome

Freitas, V. H. de (2002) *Soil Management and Conservation for Small Farms; Strategies and Methods of Introduction, Techniques and Equipment*, FAO Soils Bulletin 77, FAO, Rome

Giller, K. E., Baere, M. H., Lavelle, P., Izac, A.-M. N. and Swift, M. J. (1997) 'Agricultural intensification, soil biodiversity and ecosystem function', *Applied Soil Ecology*, vol 6, pp3–16

Giller, K. E., Bignell, D. E., Lavelle, P., Swift, M. J., Barrios, E., Moreira, F., van Noordwijk, M., Barois, I., Karanja, N. and Huising, J. (2005) 'Soil biodiversity in rapidly changing tropical landscapes: Scaling down and scaling up', in M. B. Usher, R. Bardgett and D. W. Hopkins (eds) *Biological Diversity and Function in Soils*, Cambridge University Press, Cambridge

Huising, E. J. (1993) 'Land use zones and land use patterns in the Atlantic Zone of Costa Rica. A pattern recognition approach to land use inventory at the sub-regional scale using remote sensing and GIS applying an object oriented and data-driven approach', PhD Thesis, Wageningen Agricultural University, The Netherlands

ILACO (1985) *Compendium for Agricultural Development in the Tropics and Sub-Tropics*, Elsevier Science Publishers BV, Amsterdam, The Netherlands

Iowa State University (2004) 'Weed Science', www.weeds.iastate.edu/mgmt/2004/world.shtml

Joshi, L., Wibawa, G., Vincent, G., Boutin, D., Akiefnawati, R., Manurung, G., van Noordwijk, M. and Williams, S. (2002) *Jungle Rubber: A Traditional Agroforestry System Under Pressure*, ICRAF-Southeast Asia Regional Research Programme, Bogor, Indonesia

Kuechler, A. W. and Zonneveld, I. S. (eds) (1988) *Vegetation Mapping. Handbook of Vegetation Science*, vol 10, Kluwer Academic, Dordrecht, The Netherlands

Lillesand, T. M. and Kiefer, R. W. (1987) *Remote Sensing and Image Interpretation*, John Wiley & Sons, New York

Molenaar, M. (1991) 'Formal data structures, object dynamics and consistency rules', in H. Ebner, D. Fritsch and C. Heipke (eds) *Digital Photogrammetric Systems*, Wichman Verlag, Karlsruhe

Molenaar, M. (1998) *An Introduction to the Theory of Spatial Object Modelling for GIS*, Taylor & Francis Ltd, London

Oregon State University (2005) *Pacific Northwest Weed Management Handbook*, www.ippc.orst.edu/pnw/weeds

Ruthenberg, H., MacArthur, J. D., Zandstra, H. D. and Collinsons, M. P. (1980) *Farming Systems in the Tropics*, 3rd edition, Clarendon Press, Oxford

Stocking, M. and Murnaghan, N. (2001) *Handbook for the Field Assessment of Land Degradation*, Earthscan Publications Ltd, London

Vink, A. P. A. (1975) *Land Use in Advancing Agriculture*, Springer Verlag, Berlin

WOCAT, World Overview of Conservation Approaches and Techniques, www.wocat.org/swcmeas.asp

Index

CTA

partageons les connaissances au profit des communautés rurales

sharing knowledge, improving rural livelihoods

The Technical Centre for Agricultural and Rural Cooperation (CTA) was established in 1983 under the Lomé Convention between the ACP (African, Caribbean and Pacific) Group of States and the European Union Member States. Since 2000, it has operated within the framework of the ACP-EC Cotonou Agreement.

CTA's tasks are to develop and provide services that improve access to information for agricultural and rural development, and to strengthen the capacity of ACP countries to produce, acquire, exchange and utilise information in this area. CTA's programmes are designed to: provide a wide range of information products and services and enhance awareness of relevant information sources; promote the integrated use of appropriate communication channels and intensify contacts and information exchange (particularly intra-ACP); and develop ACP capacity to generate and manage agricultural information and to formulate ICM strategies, including those relevant to science and technology. CTA's work incorporates new developments in methodologies and cross-cutting issues such as gender and social capital.

CTA is financed by the European Union.

CTA
Postbus 380
6700 AJ Wageningen
The Netherlands

Website: www.cta.int